GW00729130

Atlas of Sussex Butterflies

with a commentary
on their
changing conservation status

by

Joyce and Peter Gay

Butterfly Conservation
Sussex Branch
1996

The views expressed in this Atlas are those of the authors
and do not necessarily reflect those of
Butterfly Conservation

First published 1996
by Butterfly Conservation, Sussex Branch
Wellbrook, High Street, Henfield, Sussex, BN5 9DD
Copyright © 1996 Butterfly Conservation, Sussex Branch
Text Copyright © 1996 Joyce and Peter Gay
Front cover illustration Copyright © 1996 Thomas Barber

ISBN   0 9522602 4 7

Printed on chlorine-free paper
by Bonner and Jenkins Printing Limited of Lower Beeding, Sussex

Contents

Dedicated
to the many people who gave so freely
of their time in providing the butterfly records
that made this Atlas possible

Part 1   The survey
Aims, method and background

Introduction   The aims and scope of the survey

Recording of butterflies is not new in Sussex. In the 19th Century the building of the railways enabled people from Croydon and London to visit Sussex more easily. Many Victorian naturalists took advantage of this and kept meticulous notes on all they saw. It is interesting to read of the areas they visited and of the clouds of butterflies they saw but depressing to realise that many species have declined since then. On the other hand, the distribution of some species appears little changed -- the Brown Hairstreak is still found over much the same area of mid-Sussex as it was a century ago.

During the late 1970s a concerted effort was made by the West Sussex Recording Group to discover the distribution of all butterfly species in West Sussex. They produced maps based on the presence of each species in every 2km x 2km square (known as tetrads) of the National Grid and published their results in January 1980. Subsequently, recording for the whole of Sussex was continued by members of the Sussex branch of Butterfly Conservation, several natural history societies and interested individuals. The South Downs were reasonably covered as were parts of Ashdown Forest and the nature reserves of English Nature and the Sussex Wildlife Trust.

The origin of the survey

It became clear that there were large areas of Sussex which had no recent records, so in 1989 our society began more systematic recording with the aim of producing an atlas showing the up-to-date distribution of Sussex butterflies.   Unlike the Victorians who recorded to fulfil their natural curiosity, the principal aim of our survey has been to provide a factual basis for action to secure the conservation of butterflies in Sussex.

We have included in this Atlas only records collected in 1989-94 inclusive. A longer period or the inclusion of earlier historical records could have masked changes in distribution that may be important. A shorter time would have made it impracticable to collect enough records from such a large area. Initially the intention had been for the survey to run from 1990 to 1994, and to have a short trial of the proposed system in 1989. As the trial showed no problems requiring a change it seemed sensible to regard these relatively few 1989 records as part of the survey, but for most purposes the records can be taken as covering the five year period of 1990-94.

Why survey common species?

It is understandable that more interest will be focused on the scarcer species than on the widespread ones. Nevertheless, our survey has sought to produce as complete a record as is possible for all species, regardless of their scarcity, in order to give a full and balanced picture that can be relied on by all who wish to refer to the Atlas.

But there is an additional reason for including the common species. Whilst we have enough experience to predict the effect on most of our species of the various usual changes in land use and management, there are bound to be changes that at present cannot be foreseen and whose consequences therefore cannot be predicted. It is not just external factors that can surprise us. In the natural world, species sometimes undergo dramatic changes for no obvious or discernible reason. They have done so in the past and will doubtless do so in future. It is our hope that this Atlas will help to provide a reliable baseline against which to assess any such changes.

Scope of the Atlas

In producing this Atlas we have had in mind particularly those who enjoy looking at butterflies or photographing them and those who need the information as a help in their decisions. We have tried to avoid duplicating existing butterfly identification books (see Appendix 4 for some of these) but we have included practical tips that should make field identification easier for the interested amateur. Appendix 5 gives suggestions on when and where butterflies can be seen in Sussex.

Survey methods   Reliability of records

Our survey was based on recording the presence or absence of each species of butterfly in the 1000 or so tetrads in Sussex. The tetrad is the recording unit now used by various specialist societies, eg for birds, plants, dragonflies. It makes use of the grid lines found on Ordnance Survey maps, each tetrad being a square of 2km x 2km.

We devised a simple standard record card for use by the many recorders who provided the information without which this Atlas would not have been possible. These volunteers were mostly members of our branch or other natural history societies. All the records were transferred to computer using a program specially devised by a member of the Sussex branch of our society -- this enabled us periodically to produce maps showing progress with the survey, and so pin-point gaps where extra effort was needed. The map on page 8 summarises the number of species that the survey recorded from each tetrad. The individual species maps were produced using the Levana computer program.

Some limitations

All records were scrutinised, and any which raised questions were investigated further. The lack of any record of a species in any tetrad does not necessarily indicate its absence, and we recognise that some areas were not searched intensively enough.

To ensure that the true distribution of every species is properly recorded would ideally require a visit to each site on at least four occasions during the flight season. This was not always possible and undoubtedly there will have been some under-recording, but we tried to give extra attention to any areas where there were apparent serious gaps. North-east Sussex is one area that we know needs further work. Some areas with low numbers of recorded species genuinely represent areas of lesser value for their butterflies, for example, the coastal plain to the south-east of Chichester. The South Downs shows up as an obvious belt where in general, numbers of species are higher that the average for Sussex. One or two 'high spots' reflect some more popular areas for butterfly hunting, but some probably indicate where recorders live, a common problem with all recording schemes.

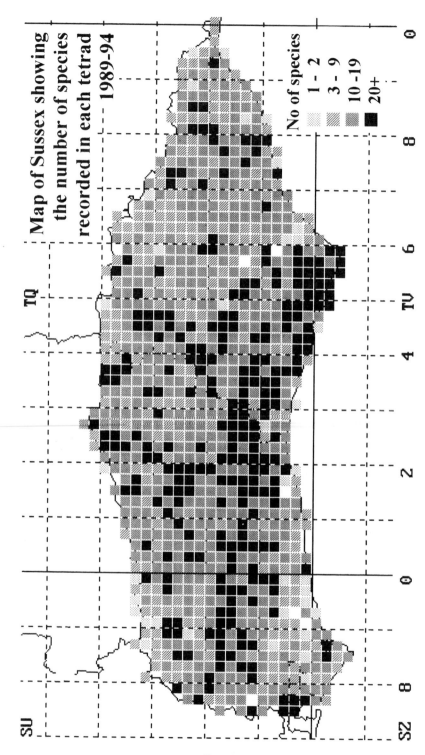

Map of Sussex showing the number of species recorded in each tetrad 1989-94

No of species
1 - 2
3 - 9
10 -19
20+

We are aware that captive bred specimens are being released into the wild and that it is seldom possible to distinguish these from naturally occurring specimens. Most of the released specimens will not survive for long, particularly if they are not released into their appropriate habitat. It does mean however that in the case of the scarcer native species the true status of a colony can be masked by the presence of released specimens, and lead to delay in taking remedial conservation measures.

Our survey was not designed to record actual numbers of butterflies, although we have made use of such data where it was available. In particular we incorporated the results from several butterfly transects that are walked weekly from April to September. These were set up by various organisations to monitor long-term changes in numbers of butterflies, and Appendix 2 gives an outline of the scheme with a selection of the transect sites.

Species distribution and frequency   The maps and what they show

The distribution of most species in Sussex is shown in this Atlas by means of maps that incorporate all the data recorded during the survey. The presence of a species in any tetrad is indicated by a spot on the relevant species map. If no spot is shown, this means no records were received for that species in that tetrad. On each map the 10km lines of the National Grid have been added to help readers to locate the exact position of any tetrad record.

Assessment of butterfly frequency

The number of tetrads in Sussex from which each species has been recorded is given in the individual species accounts immediately under the butterfly name together with a general assessment of its overall frequency. The frequency category can only be a personal value judgement and seeks to reflect both the number of tetrads from which the butterfly has been recorded and the general abundance of the species throughout the county. There are further complications, as we have had to try to describe in one or two words what is in reality a most complex situation, so the frequency category given to each species should be regarded as only an approximation.

A few examples will show some of the difficulties.

--Some species are widespread but are only found in very low numbers at any one site, eg Comma. Such species will be shown on the maps as being present in a good number of tetrads, yet the total number of individual butterflies can be relatively small.

--Conversely other species, such as the Chalkhill Blue, are represented on the map by only a small number of tetrads because they are restricted to relatively few sites. However, on each individual site they are often present in quite high numbers.

--It must also be borne in mind that numbers of most butterflies fluctuate, sometimes wildly, from year to year. In some cases this will be in response to favourable or disastrous climatic conditions, eg the numbers of Adonis Blue crashed as a result of the cold wet spring of 1993 but fortunately the following year brought a hot summer that enabled it to begin a successful recovery. Other species appear to suffer from predations when the butterfly's numbers build up -- this happens to the Holly Blue which is attacked by a parasitic wasp every three or four years producing the cyclical fluctations in the numbers which so characterises this butterfly.

The frequency categories that we have used are abundant, common, locally common, frequent, locally frequent, scarce, rare and casual. These are defined in Appendix 1, where each species is listed under its appropriate category.

Distribution ecology of Sussex butterflies

In Sussex we are fortunate in having a good number of butterfly species -- 45 out of a total of some 58 species that breed in Britain. In part this is because Sussex, being in a warm corner of the country, provides the butterflies with the sunshine they need to fly and hence to breed. Several of our butterflies are very much at the edge of their range, being essentially south-European species that can just get by in some more sheltered spots in the county. The distribution of some of our blue butterflies is further limited by their specialised relationship with ants that also require the warm conditions that we have in Sussex.

Three of our 45 species are on our list simply because we are close to the Continent. These arrive every year on migration, and several other species, such as the Monarch, are seen in ones and twos in some years.

Within Sussex, butterfly distribution often reflects the underlying geology -- see the outline geological map on page 12. This is because the distribution of any butterfly largely depends on the presence of its caterpillar food-plant, which is itself often limited by the geology and its related soils. A cursory examination of the distribution maps of such species as the Adonis Blue will show that it is restricted to the Chalk. The Silver-studded Blue is one of those species that in Sussex is confined to the sandy soils of the Lower Greensand and various zones of the Hastings Beds.

The distribution of the Marbled White is less straightforward. Whilst being found mostly on the Chalk and usually thought of as a downland species, it occurs also as occasional colonies in the Weald. It is essentially a species of unimproved basic and neutral grassland and is not so dependent on the underlying geology as appears at first sight.

The Small Tortoiseshell is found throughout the county because its caterpillar food-plant is the nettle. By contrast, the Green Hairstreak is able to occur across a wide area for an entirely different reason -- the caterpillar feeds on different food-plants on the Downs (mostly rock-rose) compared to those it uses elsewhere (mainly gorse).

Anyone interested in examining the relationship of butterfly distribution in Sussex to their caterpillar food-plants will find it useful to consult the Sussex Plant Atlas and its Supplement (see Appendix 4). This gives further information on the relationship between many plants and their underlying geology and soils. That Atlas also uses the same tetrad-based maps as we do in this butterfly Atlas, making direct comparison straightforward.

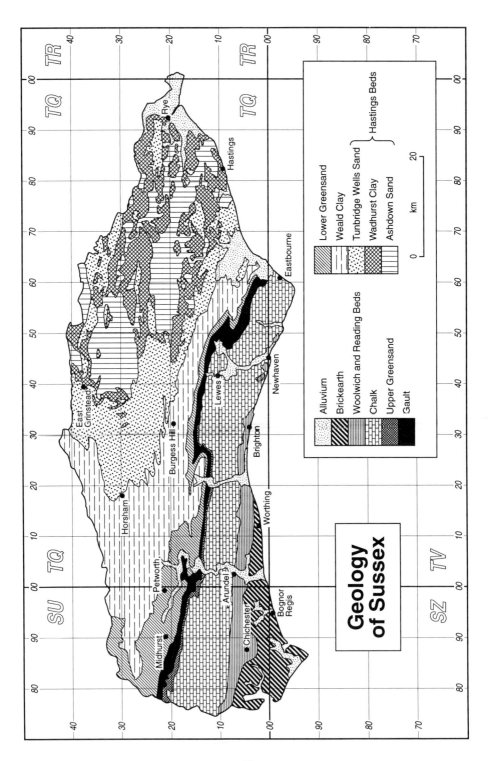

Geology of Sussex

Legend:

Alluvium
Brickearth
Woolwich and Reading Beds
Chalk
Upper Greensand
Gault

Lower Greensand
Weald Clay
Tunbridge Wells Sand } Hastings Beds
Wadhurst Clay
Ashdown Sand

km

0   20

Locations: Rye, Hastings, Eastbourne, Newhaven, Lewes, Brighton, Worthing, Burgess Hill, East Grinstead, Horsham, Petworth, Midhurst, Arundel, Chichester, Bognor Regis

Past changes in the countryside

Extinctions

During the 1970s some 47 or 48 species were thought to breed regularly in Sussex. Of these the Large Tortoiseshell and High Brown Fritillary have not been reliably recorded as breeding in the wild since the mid 1980s despite special searches for them. It is known that captive-bred specimens, especially of the Large Tortoiseshell, have been released into the wild at several places. There is no record of these releases having resulted in either of these species becoming re-established in the wild, and both are now regarded as being extinct in Sussex, although any evidence to the contrary should continue to be investigated.

Downland changes

Throughout the 1970s and 80s, several species declined at an alarming rate both in numbers and the extent of their distribution. This is undoubtedly the result of loss of suitable habitats mainly because of changes in the type or intensity of farming practices. The loss of old downland grassland by conversion to cereals had been a problem more of the late 1950s and early 1960s. This eventually led to equally profound changes on the downland areas that had not been ploughed because it became uneconomic to graze them.

This general lack of grazing on the Downs resulted in scrub encroachment and the development of a coarse grass sward on such grasslands that remained. This eliminated the very short sward with its occasional bare patches that several downland butterfly species need, and progressively choked out their foodplants. Species most drastically affected by these changes include the Silver-spotted Skipper, Adonis Blue, Duke of Burgundy, Dark Green Fritillary, and Grayling. Other species similarly affected but rather less disastrously are Dingy and Grizzled Skippers, Small Blue, Brown Argus, and Chalkhill Blue.

Spraying, and the ploughing of field edges have had adverse effects throughout the county, but were more significant for butterflies in the Weald and the coastal plain. By removing the important remnants of a whole range of wild plants that supported the more common and widespread butterflies (the species familiar to everyone) these practices undoubtedly contributed to the general decline in the overall numbers of butterflies to be seen through the countryside.

Changes in woodland management

There have been equally significant changes in the county's woodlands. As it became less economic to work coppices they became overgrown, suppressing the violets and other flowers that are such an important feature of the early stages of the regular coppice cycle. It is these flower-rich areas of extensive coppice woodland that are the life-blood of such species as the Wood White and the Small Pearl-bordered and Pearl-bordered Fritillaries.

Likewise the conversion of coppices and other woodland to more economic conifer crops was too great a change for any of our woodland butterflies to withstand. Although the rides retain some value for butterflies, this is only temporary -- as the trees grow, the rides become too shady either for the butterflies themselves or for the flowers on which they depend. Such changes have eliminated the Duke of Burgundy from its last principal woodland site away from the Downs.

Some improvements

However, the various changes were not wholly bad. Most notable was the improvement in numbers of the White Admiral. This species requires shaded woodland where there is an abundance of its caterpillar food-plant, old honeysuckle with its long trailing strands. These conditions were produced by the general lack of woodland management, yet in the case of the other species mentioned earlier it was increased shading that caused their decline. This demonstrates the difficulty of finding a balance that suits every species.

A major change, this time a natural one, was the havoc created in many Sussex woodlands by the great storms of 1987 and 1990, the second being early in our survey. The great gaps torn in what had been a closed woodland canopy, created the open woodland conditions that in recent historical times had been produced by coppicing and felling. The Grizzled Skipper was the species which appears to have benefited most greatly, although it took a year or so for the effects to show fully. This opening of the canopy has been to the disadvantage of the White Admiral, putting into reverse the advantages it had received from the earlier neglect of woodlands. As these woodland gaps fill naturally or are replanted, it can be expected that the Grizzled Skipper will decrease in these habitats.

Recent and current changes in the countryside

Nature reserves and other protective ownerships
Over recent years the various conservation organisations including the National Trust have added significantly to their holdings in Sussex, and most of these are of value for butterflies even though in most cases this will not have been the reason for the acquisition. Although our society only owns and manages one small reserve in Sussex, the other nature conservation bodies do take butterfly conservation into account on their reserves. The County and District Councils are also increasingly taking on a conservation role in managing their growing number of local nature reserves.

But nature reserves alone are not sufficient if butterflies are to have a secure future or are to be seen as widely as they once were. What is needed is for butterfly and other wildlife conservation to be an integral part of normal land management, and there are indications that this is beginning to happen.

The new changes in the countryside   Cash for conservation
Our survey was undertaken at a time when several major changes affecting management in the countryside were beginning. If they are imaginatively handled they all offer considerable scope for significant benefit to butterflies. The schemes concerned are the Environmentally Sensitive Areas (ESA) scheme for the Sussex Downs (administered by MAFF), the Countryside Stewardship scheme (Countryside Commission), and the renewed interest in coppicing by both the Forest Authority and Forest Enterprise -- both formerly parts of the Forestry Commission. By making funds available from central government, all these schemes are giving the necessary financial incentive to re-introduce the more traditional types of management that are favourable to butterflies.

Many areas of the Sussex Downs that had for years been under the plough are now being brought back to grassland particularly as a result of the funds available through the ESA. Although it will take years for the floristically rich downland turf favoured by butterflies to return, the re-introduction of more traditional grazing management is undoubtedly a major step in the right direction. Several Countryside Stewardship schemes are similarly providing the incentive for improved management for wildlife, including butterflies. Sites where this scheme has been introduced are scattered through Sussex and normally include arrangements for public access, so that as well as improving the habitat for butterflies they are also places where people can take pleasure in seeing more butterflies in the wild.

The Forest Authority is able to give grants to owners of private woodlands for the management of coppice, and undoubtedly fuller use could be made of this for conservation purposes.

A problematical scheme

Another change on farmland but one that is more problematical, is the result of the Set Aside scheme (also administered through MAFF with central government funds). Whilst this does indeed result in a wealth of weeds where butterflies can nectar, there is not the continuity of habitat that most butterflies need if they are to breed successfully. It is a butterfly's ability to breed on an area that must be the main criterion in assessing whether any scheme is of real benefit to it.

Strategies for nature conservation

Overall policies for nature conservation have been adopted in recent years by both East and West County Councils. It is hoped that this action will have pushed conservation higher up the agenda of all local authorities in Sussex, but it will not have removed the need for promoting conservation interests and fighting individual cases, as inevitably there will be differences because of the many conflicting demands of a modern society.

The Sussex Downs Conservation Board is currently developing a conservation management strategy that will provide the opportunity for the special needs of wildlife conservation to be recognised throughout this important area. The success of the strategy will of course depend on the degree to which it is adopted by the various local authorities represented on the Board.

There have also been important moves towards the formal recognition of sites that are of special importance for their wildlife but do not qualify as SSSIs. Over the last few years the County Councils and District Councils, with the specialist support of wildlife bodies including our own society, have been identifying Sites of Nature Conservation Importance (SNCIs). This will help to ensure that these important sites are known both to their owners and public bodies, and are not lost or damaged by accident.

Management for conservation

Forest Enterprise are already managing a number of their coppice areas in more traditional ways, one of those areas, Rewell Wood near Arundel, being a very important site for the Pearl-bordered Fritillary.

Several local authorities in recent years have also been taking part in the practical side of nature conservation. Both the East and West Sussex County Councils are increasing their interest in the management of selected roadside verges for their wildlife, many of them being of major significance as reservoirs for a very wide range of butterfly species. Unfortunately there is cause for concern at the spreading of soil on roadside verges where, for example, several good colonies of the Orange Tip have been eliminated in recent years as a result of what seems to be a growing practice.

The Ranger Service of the Sussex Downs Conservation Board is playing an increasingly active role in helping with the management of some important butterfly sites.

A welcome change that has scope for benefiting butterflies in all parts of the county, both in town and countryside, is the increasing recognition that churchyards provide an excellent opportunity for wildlife conservation. The grassland in most churchyards is quite ancient, with a rich variety of plant species. If mowing is delayed until after these plants have flowered, they provide a good nectar source for butterflies in a place where many people can take pleasure in seeing them.

Pointers for the future

After some 50 years of continued decline of butterflies in Sussex, the changes in the countryside that led to this seem at last to be showing signs of moving in the opposite direction. This new trend appears to stem principally from the changes in the economics of farming and the greatly heightened concern for the environment. Provided the various new schemes and policies for more sensitive management of the countryside are continued and imaginatively extended, there is now reason for cautious optimism for the future of butterflies in Sussex.

The progress in butterfly conservation in Sussex made in recent years is largely the result of a number of partnerships. Our branch has worked closely with a whole range of public bodies and with many individual landowners and land managers. Our society has provided the expertise, our volunteers have provided their time and physical effort, public bodies and charities have supported us with the necessary cash, and landowners have been generous with permission to carry out the conservation tasks on their land. Partnerships of this kind will surely be even more relevant in the future.

But despite the improvements, there are practical problems that call for action. The four rare species that have reached critical levels are Silver-spotted Skipper, Wood White, Duke of Burgundy and Grayling. The Sussex branch of Butterfly Conservation is preparing priority Action Plans for these in co-operation with its neighbouring branches.

The Small Pearl-bordered and Pearl-bordered Fritillaries are also in the priority list, both requiring a major expansion in active coppice management. Success here can only be achieved if woodland owners consider that the financial returns are worthwhile, and this will probably depend largely on the scale of the available incentives.

The way forward for the Dark Green Fritillary is much less clear, as the reason for its recent drop in numbers is not understood. This is a case where further research is probably essential, but meanwhile it would be prudent to closely monitor its status and encourage continued light grazing on its main sites.

For the downland butterflies in general, the key continues to be adequate grazing, but the precise level of grazing on individual sites can be critical for species like the Adonis Blue, and this should be watched so that advice can be give, particularly in connection with the ESA schemes.

We shall continue to need up-to-date facts about the distribution of Sussex butterflies. We are already continuing the tetrad recording scheme -- it is a practical way of identifying any changes in the general status of Sussex butterflies requiring remedial measures. We have also expanded the scheme so that we can assemble more detailed records on the more important sites identified during the 1989-94 survey. This addition will be particularly valuable in deciding how these sites should be managed. These records will also be incorporated into the national recording scheme that Butterfly Conservation is planning as a Millennium project that will lead in the year 2000 to a butterfly Atlas for the whole of Britain.

The challenge

The theme of this book has been to describe the present distribution and status of Sussex butterflies in the light of their ecology and against a background of recent and current changes in their habitat. The principal purpose has been to provide a factual basis for their conservation, and we have made some suggestions for future action, for without action by the many parties concerned, conservation cannot succeed. We have suggested that the key to effective action is partnership, and we are convinced that volunteers from all walks of life will continue to have as important a role in the future as they have done in the past.

This Atlas has provided the basic facts about the present distribution of the butterflies of Sussex. The challenge now is to take action so that our butterflies can once more become a conspicuous part of the Sussex scene in town and countryside.

Part 2   The butterflies of Sussex Distribution maps and outline of their biology

Small Skipper

Common   525 tetrads

Thymelicus sylvestris

This small orange-coloured butterfly with its quick whirring flight can be seen from June to August. The male manoeuvres skilfully through tall grass, and then in typical skipper fashion makes sudden dashes (presumably the origin ot the name 'skipper') and apparently disappears. The female is much more sedentary. At rest both normally adopt the typical 'skipper position' that is described for the Large Skipper.

The orange under-tip of its antennae distinguishes it from the almost identical Essex Skipper. This feature can be seen if the butterfly is approached closely, but this calls for some patience.

The Small Skipper is found in almost every patch of rough grassland, in town and country. It is not common if the grass is kept short by grazing or continuous mowing, and colonies have undoubtedly been lost due to ploughing, intensive agriculture and excessive tidiness. All of these tend to suppress the caterpillar's foodplant, Yorkshire fog. It can be seen in large numbers nectaring on thistles, knapweed and many other flowers.

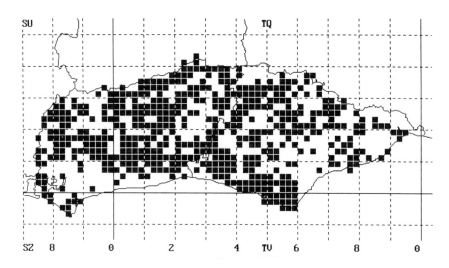

Essex Skipper

Thymelicus lineola

Common   464 tetrads

The tawny orange Essex Skipper is found from late June to the end of August, emerging about ten days after the Small Skipper. These two species occupy similar habitats and often fly together, although one or other is usually the dominant.

Because of the close similarity of these two species the Essex Skipper is probably under-recorded. Identification in the field is not easy as it is necessary to look for the jet-black under-tips of its antennae. Considerable effort was made during our survey to identify all specimens, and we suspect that both species are present in virtually all areas where one or other species was recorded.

The main difference between these two species is unfortunately something that cannot be seen by examining the adult butterfly. The Essex Skipper overwinters as an egg, whereas the Small Skipper spends the winter as a caterpillar.

The Essex Skipper is found throughout both East and West Sussex on wide roadside verges, downland and across the Weald, wherever the caterpillar foodplants such as cocksfoot and creeping soft-grass grow in tall undisturbed grassland. The adults can easily be found nectaring on such flowers as knapweed, thistles and ragwort.

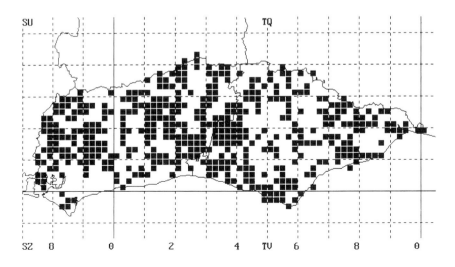

Silver-spotted Skipper

Hesperia comma

Rare   9 tetrads

   This is a butterfly restricted to a very few sites in East Sussex, although it was found in West Sussex up to the 1970s. The downland of one such West Sussex site was ploughed and is now under cereals. Although the larval foodplant, sheep's fescue, still grows on the edges of the track alongside this former site, it is unlikely that the species will re-appear here.

   The Silver-spotted Skipper needs very sheltered sites to be successful as it is one of several species on the northern edge of their European range. It likes small bare patches of warm soil where the adults can bask in the sunshine.   The eggs are laid on small plants of sheep's fescue which overhang these patches of warm bare ground.   The sites in East Sussex where this species occurs are usually cattle grazed during the winter, which is maintaining the correct habitat.

   The butterfly flies in late July and August, and has a fast whirring flight close to the ground. At rest it is possible to see the silver patches on the greeny-brown underside of the wings that distinguish it from the Large Skipper which has no silver on its chequered pattern..

   It occurs in modest numbers on National Trust property near Alfriston, but to be certain of seeing them in flight choose a very warm day.

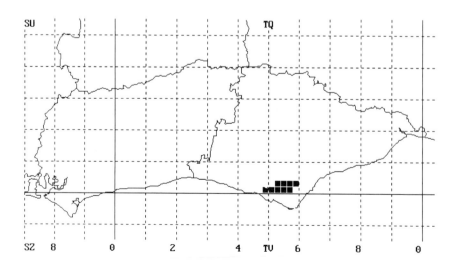

22

Large Skipper

Ochlodes venata

Common   481 tetrads

On the wing in early June to late August, this sturdy orange and brown skipper is most often seen singly, basking in the sunlight on a sheltered hedge. When at rest it usually keeps its forewings half closed, and its hindwings out flat -- in what is known as the 'skipper position'. This makes it look rather like many moths, so that it is sometimes overlooked.

It can be distinguished from the Small and Essex Skippers (which also adopt the 'skipper posistion') by the slight chequered appearance of both sides of its wings. Sometimes this chequering has led to its being wrongly reported as the Silver-spotted Skipper, which is extremely rare in our area.

The Large Skipper breeds mainly on cocksfoot grass where this is growing in an unmown, warm sheltered spot. In Sussex you are likely to find it in almost any suitable woodland edge, on hedges or downland scrub. During 1993 it was seen in abundance (up to 20 individuals in a 50 yard stretch of hedge) in the mid Sussex area.

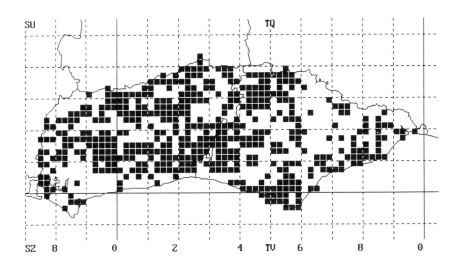

Dingy Skipper
Erynnis tages

Locally frequent   112 tetrads

    The Dingy Skipper is a small grey-brown butterfly with a swift whirring flight. Its unattractive-sounding name belies the delicate patterning of the upper surface of its wings, with a row of small white dots just inside the brownish fringe. In contrast, the superficially similar Grizzled Skipper has much bolder markings with distinctive white fringes crossed by dark lines.

    Since the 1970s many colonies have undoubtedly been lost due to more intensive cultivation and the ploughing of field edges. Its main surviving colonies are concentrated on the Downs, with a much smaller number of sites scattered across the Weald, in woods, field edges and in rough grasslands. It is however almost certainly under-recorded because it looks so much like a moth (especially the day-flying Mother Shipton) and it is worth searching for in any area that has a large amount of its larval foodplant, bird's-foot trefoil. .

    The Dingy Skipper is one of the skippers that at rest holds it wings flat like many moths, and does not adopt the 'skipper position' that is described for the Large Skipper. Its main flight period is May to June, with a small second brood recorded in August during hot years such as 1990.

    On the Downs it can often be seen on the bare chalky sides of tracks where it basks in the sun with its wings outstretched.

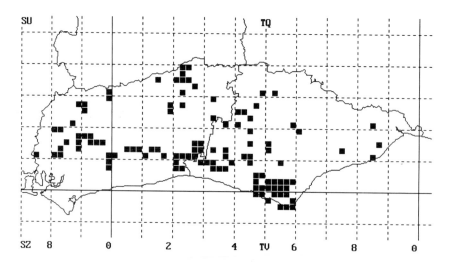

Grizzled Skipper
Pyrgus malvae

Locally frequent   111 tetrads

    A small dark-coloured butterfly with white irregular spots on the upper surface, and a black and white chequered fringe to the wings. It is found in small colonies all along the Downs and can be seen in similar situations as the Dingy Skipper, from as early as late April until mid June. It prefers sites where there are patches of bare ground that reflect the sun's warmth on to the plants where the female lays her eggs, and can most easily be seen when it basks in the sun with its wings outstretched. Like the Dingy Skipper it does not adopt the 'skipper position'.

    It is also locally abundant in open areas of some of the larger woods in East Sussex where the 1987 hurricane opened up the canopy. These areas allow in more sunlight which favours the growth of the larval foodplants, wild strawberry and creeping cinquefoil. As these areas are replanted, doubtless the numbers of Grizzled Skippers will decrease, and once more they will be confined to the rides and woodland edges.

    Away from the Downs this is another under-recorded species, and it is worth searching throughout the Weald for undiscovered colonies.

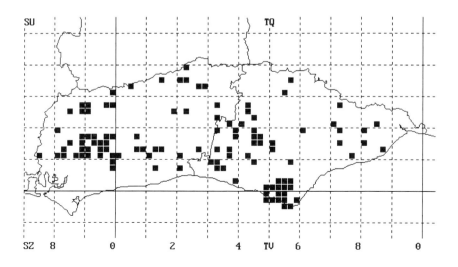

25

Wood White
Leptidea   sinapis

Rare   4 tetrads

   This small fragile-looking butterfly is now confined to a small area of West Sussex mainly along the Surrey borders. It can be recognised by its weak flight with slow wing-flaps which make it possible to see the large greyish apical spots on the forewing. It is on the wing mainly in June, and in good summers there may be a small second brood during August.

   The distribution and numbers of this butterfly have decreased alarmingly during the last 15 years. In West Sussex it is primarily a butterfly of coppiced woodland, finding its larval foodplants (bitter vetch, meadow vetchling or bird`s-foot trefoil) alongside damp woodland rides and ditches. Much of its original habitat has now been put down to conifers or has become too overgrown and shady for it to survive. Recorders during the 1994 season reported only three individual butterflies, all in one tetrad.

   Our society recently commisioned an investigation of this species in north-west Sussex, which enabled us to make proposals for the necessary conservation management to aid the species in its recovery. On the basis of this, Forest Enterprise, West Sussex County Council and English Nature have begun conservation work in certain woods to provide the more open conditions that the butterfly requires.

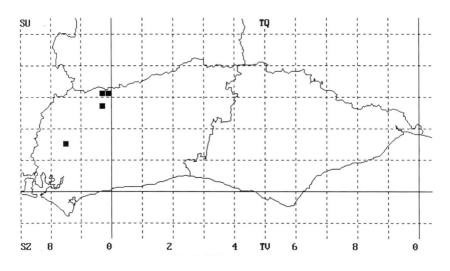

Clouded Yellow
Colias croceus

Casual   171 tetrads

These bright yellow butterflies are found most frequently near the coast, because they come to us from across the Channel and the number we get reflects their breeding success in the Mediterranean region. A few individuals of this migratory species arrive in most years but 1992 and 1994 saw a very good influx so that it was recorded widely throughout Sussex. In these 'good' years many were seen in clover fields, pastures and nectaring in gardens. Some were found to be breeding on lucerne and clover near the coast around Eastbourne. However, despite the mild winters there have been no reports of over-wintering individuals.

They fly very fast, and often are seen only as a yellowy-orange flash. This is usually the only time that the orange of the upper wing surface can be seen, as at rest they always fold their wings and then appear bright yellow. A small proportion of females have a pale upper wing surface (*helice*), some of which have probably led to erroneous reports of the much rarer Pale Clouded Yellow.

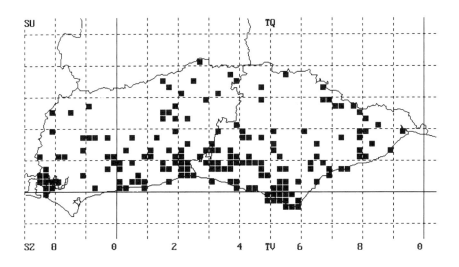

Brimstone

Gonepteryx rhamni

Common 483 tetrads

A widespread butterfly that emerges from hibernation with the first warm days of spring. The male in particular is easily spotted because of its large size and yellow colour, whereas the female is much paler and in flight can easily be confused with a Large White. When at rest the Brimstone's characteristic `hooked` wing shape immediately distinguishes it from all other 'whites'.

The butterfly can be seen during spring and again from August until late autumn especially on the Downs and in many woods of both East and West Sussex wherever the larval foodplants, either purging buckthorn or alder buckthorn, grow. They usually fly singly, but especially during late summer they congregate when feeding from thistles and other flowers.

The caterpillar is surpringly easy to spot despite having the same green colour as the leaves it feeds on. Its presence is revealed by the nibbles in the young leaves and it can be seen resting along the midrib on the upper surface of the leaf.

Good localities to look for the Brimstone are along the rides in Houghton Forest, West Sussex and around the Vert Wood area near East Hoathly in East Sussex.

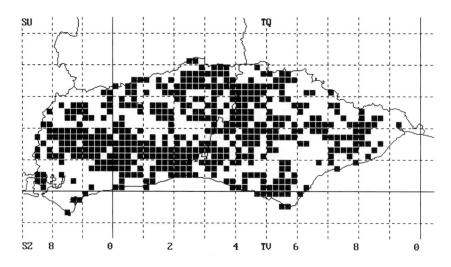

Large White

Abundant   854 tetrads

Pieris brassicae

This butterfly, often called the Cabbage White, needs little introduction, especially to gardeners because it is a pest of brassicas. The top surface of both the male and female is very white with large black tips extending about half way down the outside edge of the forewing. The female also has two conspicuous black spots on the forewings that can easily be seen in flight. It lays its yellow eggs in large batches which develop into black and green patterned, hairy caterpillars.

It is easy to confuse the Large with the Small White butterfly. The most certain distinguishing features are that the Small White has much smaller black markings on its front wing-tips and lays its pale yellow eggs singly.

Our numbers of Large Whites are greatly increased by mass migrations from across the Channel, and if these do not take place it is seen in much lower numbers. In 1992 the air on the Downs behind Brighton resembled an early snow storm -- thousands of Large Whites had arrived on the southerly winds.

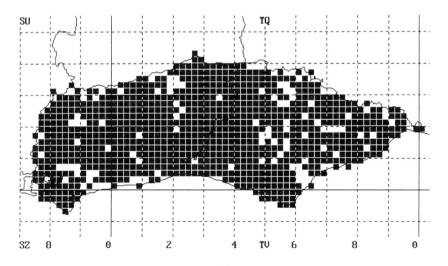

Small White

Pieris rapae

Abundant 773 tetrads

    This medium-sized white butterfly can be seen almost everywhere throughout both counties during spring and summer, being particularly abundant where there are brassica crops. The numbers vary widely from one year to the next and can be greatly increased by migrations from the Continent.

    It is easy to confuse it with the Large White, but the Small White is normally smaller and the black tip to its upper fore-wings extends only a short way down the outside edge. The eggs are pale yellow and are laid singly (not in batches like the Large White) on cabbage or nasturtium leaves in gardens, and in the wild on crucifers such as garlic mustard. The caterpillars are grass-green which makes them well camouflaged on cabbage and other leaves, and they do not have the marked patterning of the Large White caterpillars.

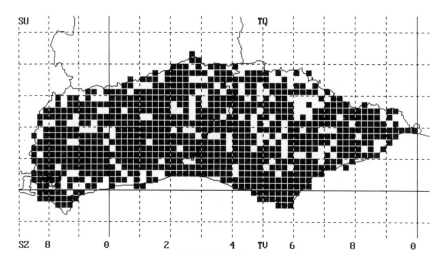

Green-veined White

Common   582 tetrads

Pieris napi

    This attractive butterfly should not be grouped as a `cabbage pest` with the Large and Small Whites as it lays its eggs on wild Crucifers such as garlic mustard or hedge mustard. Its fluttering flight and grey-green veining on the undersides of the hind wings are its main distinguishing characteristics, but its upper wing-surface is generally similar to that of both the Large and Small Whites. The green veins of the second brood are much fainter so from July onwards more care is needed to identify this species.

    It can be seen flying along damp hedge-sides, woodland rides and coppiced areas in April but is more common in July and August. It is then an occasional visitor to gardens where it sometimes lays its eggs on nasturtium leaves. Any 'white' seen in woodland is most likely to be this species.

    Although probably under-recorded it occurs throughout the two counties but is more likely to be seen in the Weald than on the Downs.

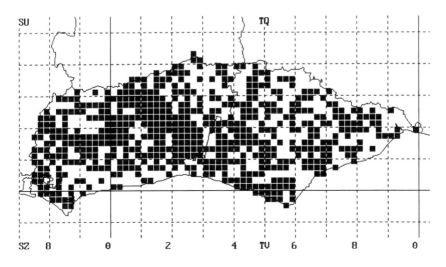

Orange Tip

Anthocharis cardamines

Common   520 tetrads

Often called the harbinger of spring, the Orange Tip is a welcome sight during late April to June. The males with their orange wing-tips are easily spotted flying along roadsides, in woods and over damp fields. During the survey it has been noticed that these butterflies emerge about a week to ten days earlier in the warmer coastal strip than those in the higher land of the High Weald. The females are more difficult to spot, often being confused with the Green-veined or Small Whites, but the grey-green mottled under-side of the hind wings is a very distinctive guide.. The eggs which are laid singly near the flower-buds of the food-plant, eg lady's smock or garlic mustard, soon turn a bright orange, making them easy to find.

This butterfly is still found throughout Sussex but undoubtedly the numbers have decreased as fields are drained, roadside verges tidied up or grassland mowed at the wrong time, thus eliminating the food-plants. Churchyards that are managed for wildlife are often good places to see this species.

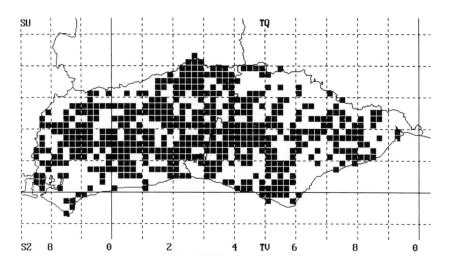

Green Hairstreak

Callophrys rubi

Locally frequent   73 tetrads

This is our only truly green butterfly and can often be seen on the Downs during early May. It occurs in low numbers scattered over much of both East and West Sussex. When flying, it is difficult to identify as it looks like a blurred brown butterfly. When resting the bright green underwings are unmistakable although amongst green vegetation it can be well camouflaged. The hairstreak (the line across the underside hind-wing that characterises this group of butterflies) is reduced to a row of white dashes. The male perches on a prominent leaf or twig and maintains his territory around this, returning again and again to the same place.

It has a wide range of plants on which it lays its eggs -- gorse, rock-rose, bird's-foot trefoil, purging and alder buckthorn or dogwood. In our area this enables the species to range widely from calcareous downland to acid heaths and over the heavier soils of the Weald. During 1993 it did very well and was recorded in 19 more tetrads than before. The numbers in each colony were also up, with several individuals being seen at once. With more intensive searching at the right time we hope to establish a more accurate picture of the distribution of this attractive butterfly.

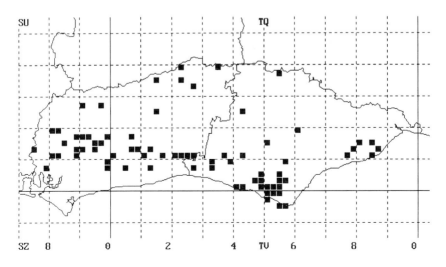

Brown Hairstreak
Thecla betulae

Scarce　　　　　　　9 tetrads

In spite of intensive searching in East Sussex, the Brown Hairstreak was found only in West Sussex, mainly on the heavy soils of the Low Weald where there are many blackthorn hedges. This is the only species whose distribution was investigated by searching for eggs during the winter months. From this it appears that in West Sussex there is only a sparse and very scattered population of this butterfly. During our survey, apart from a few places on the Downs Link footpath, there have been only seven sightings of adults, all completely by chance.

From all this it is not yet possible to conclude the true extent of the breeding population except that it is probably concentrated around the Copsale area, but this is too vague to warrant including a distribution map for this species. The number of tetrads quoted above is for the sightings of the butterfly.

The eggs were usually found at about 3 feet (1 m) from the ground on the young vigorous blackthorn suckers, but in some areas they were on larger twigs below the hedge trimming line. They are usually laid singly in the angle of the young spines and with practice can easily be seen as a gleaming white pin-head against the dark bark.

The mechanical cutting of hedgerows in winter does undoubtedly destroy many eggs, but the butterfly seems able to withstand this provided that not all hedges in an area are cut at the same time. Even better for the conservation of this species would be for the blackthorn hedges to be cut in July. By then the caterpillars will have moved to ground level to pupate and the eggs will not yet have been laid on the twigs.

The butterflies emerge in late July and into August, and both the male and female Brown Hairstreak then spend most of their time in the tree-tops, sunning themselves and feeding from the honeydew on the leaves. It is only the female that is seen at close quarters because it comes down to lay its eggs. Then it is possible to see the unmistakable golden undersurface with its hairstreak (the conspicuous white line) and its prominent tails on the hindwings. Very occasionally the females can be seen on low branches, basking in the sun with their wings outstretched.

In flight, the Gatekeeper and Vapourer moth can sometimes be confused with the Brown Hairstreak because their colours are so similar, and both fly around tree-tops. The Gatekeeper does not have the erratic, jerky movement so characteristic of all hairstreaks, and the Vapourer, whilst having a darting flight, is much smaller.

Purple Hairstreak

Quercusia quercus

Frequent   165 tetrads

    This butterfly is recorded mainly from the Low and High Weald where its larval food-plant, oak, is abundant. It is a difficult species to spot because it spends most of its time in the tree-tops, but if the sunny side of an oak tree is studied late on a July day you can often see a slight movement of a grey-looking butterfly. Sharp eyes or binoculars will then reveal one or more resting Purple Hairstreaks.

    The underside is a dove grey with the typical white line of the 'hairstreaks'. When it flies the upper wing-surface, especially that of the male, can be seen to glint purple in the sunshine, and occasionally during hot years this can also be seen when it basks at low levels with its wings outstretched. In particularly hot spells, as in 1990, they may come down to nectar on blackberry or other flowers.

    The eggs are laid singly and can be found at the base of the cluster of buds at the ends of the twigs -- they are just under 1mm in diameter, resembling tiny sea-urchins. It is worth looking for them on any oak branches recently blown down.

    East Sussex is probably under-recorded for this species as it no doubt occurs wherever there are oak trees.

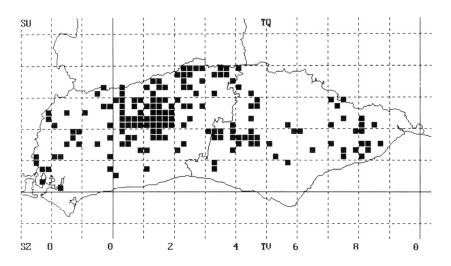

White-letter Hairstreak
Strymonidia w-album

Scarce   36 tetrads

This elusive hairstreak is found on elm, which has greatly decreased since the advent of dutch elm disease in the 1970s, although East Sussex has managed to keep a disease-free zone where large elm trees have survived. In the rest of East and West Sussex many elm suckers are now reaching 30 ft before succumbing again. Members of our society have searched these trees and the more disease-resistant wych elms for the eggs, larvae and adults of the White-letter Hairstreak. Extra colonies have already been discovered, bringing the total from 22 in 1993 up to 36 in 1994.

The butterfly can best be seen using binoculars to look for any movement of a small brown butterfly high up in the sunny canopy. In hot summers the butterfly sometimes comes down to nectar, when the white W-shaped hairstreak and the orange and black band can easily be seen on the underside of the hindwings. The eggs look like grey-brown pinhead-sized flying saucers and are laid singly often on the growth ring near the tip of the twigs, and are most easily found after the leaves have fallen. The caterpillars make large oval holes between the leaf-veins in April and May, and if this is spotted it is worth looking there for the adult in July.

It was thought that the larvae feed only on elm flowers, but White-letter Hairstreaks are surviving in Sussex on elms too small to flower. New colonies are being discovered throughout both counties except in the north-east where there are few elms, but it is almost certainly still under-recorded.

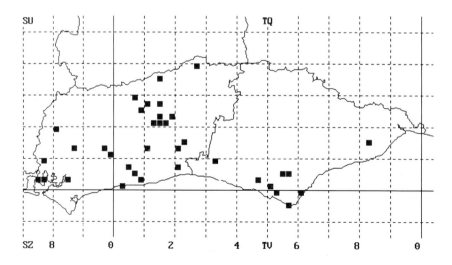

Small Copper

Lycaena phlaeas

Frequent   416 tetrads

    This small active butterfly has two broods, or even three in a hot year, but its numbers have been decreasing throughout our survey. At the beginning they could often be found in tens on some rough grassland sites, especially during August, whereas by 1994 they were much scarcer. However, numbers of all butterflies do fluctuate naturally, and we must hope that this species will benefit from the good autumn of 1994.

    The caterpillar food-plant of the Small Copper is common sorrel or sheep's sorrel, and the butterfly is most likely to be seen wherever these plants are found, especially in unimproved grasslands. Good places to see this species include the Downs, areas of rough grassland eg Oreham Common, near Henfield, West Sussex, and churchyards which have not been closely mown.

    During August the second brood individuals often nectar on fleabane, when a flash of bright coppery-orange can be seen as they dash from flower to flower. When the wings are closed they are more difficult to spot as the colour of the hindwings is a soft muted fawn.

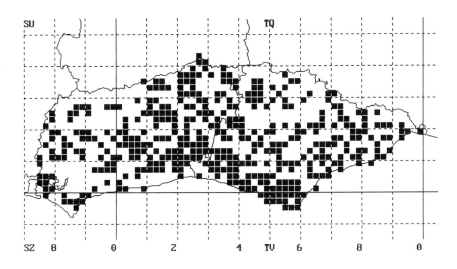

Small Blue
Scarce   64 tetrads

Cupido minimus

    In our area this small butterfly is almost completely confined to the Downs where its larval foodplant, kidney vetch, grows. It needs shelter and warmth, and usually lives in small colonies on south-facing banks. However, it can colonise much larger areas if the foodplant is present, eg to the east of the River Adur along the cutting of the new A27 where kidney vetch (unfortunately a foreign strain) had been sown by the highway authorities.

    As Britain's smallest butterfly the Small Blue is easily overlooked, and many additional colonies have been discovered during our survey period. There are fewer records from the western end of the Downs as kidney vetch is not as common there as it is to the east of the River Adur.

    The Small Blue is mainly to be seen in June, when the males can be found basking in the sunshine on long grass with wings outstretched, awaiting passing females. The underside is a distinctive silver-grey with black dots, similar to the Holly Blue, but its smaller size and dark brown of the upper surface distinguishes it from all other blues in this country.

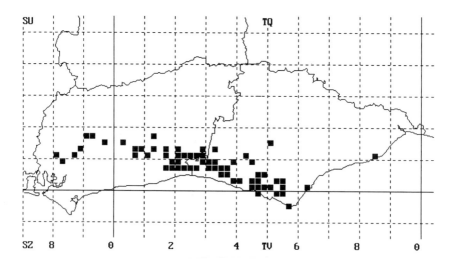

38

Silver-studded Blue
Plebejus argus

Locally frequent   27 tetrads

The male Silver-studded Blue has a lavender-blue upper surface with a wide black border, whilst the female is brown with marginal orange crescents on both wings, both being superficially similar to the Common Blue. The most easily seen diagnostic feature of the Silver-studded Blue is the row of turquoise-blue `studs` on the lower surface of the hind wing. No other British blue butterfly has this character.

The stronghold of the Silver-studded Blue is on the heathland of Ashdown Forest (including the Sussex Wildlife Trust`s reserve at Old Lodge) and Chailey Common in East Sussex, and Iping and Stedham Commons on the Greensand of West Sussex. Other smaller colonies have died out since 1980. The caterpillar feeds on young, short shoots of the heathers, particularly where these overhang bare sandy areas that reflect back the sun's heat, and also sometimes on gorse. Conservation work to recreate these conditions has been carried out by both County Councils, the Sussex Wildlife Trust and English Nature. This has obviously been successful as the Silver-studded Blue increased dramatically during 1993 and 1994 on the sites being specially managed. The colonies on Ashdown Forest appear to be responding satisfactorily to the continuing heathland management by the Conservators of the Forest.

The adult butterfly emerges in the south of the counties in mid June, whereas those on Ashdown Forest are about a fortnight later perhaps due to differences in altitude and hence temperature.

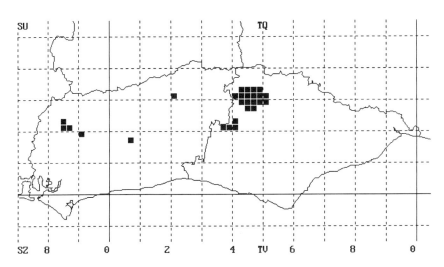

Brown Argus

Aricia agestis

Locally frequent   121 tetrads

    The male and female of this species are very similar, with deep chocolate-brown uppersides and an orange band towards the outer margins of both wings. The female is usually slightly larger, and the orange markings on the upper wing surface are more conspicuous. The underside of both is similar to that of the female Common Blue -- the differences are not easy to spot in the field, but a useful guide is that the Brown Argus flits quickly and erratically from place to place in contrast to the steadier pace of the Common Blue. The male Brown Argus perches on a flower or the end of a twig from where it defends its territory, and this is the easiest time to see it clearly. Any 'female blue' behaving aggressively to other passing butterflies is most likely to be a male Brown Argus.

    Butterflies of the first brood can be found in early May all along the Downs in the vicinity of their larval foodplant, rockrose, but it is easier to find it in July and August as the second brood usually has higher numbers. A few colonies have been found off the Downs where the caterpillars live on dove`s-foot cranesbill or common storksbill. Because of the similarity to the female Common Blue this species is almost certainly under-recorded.

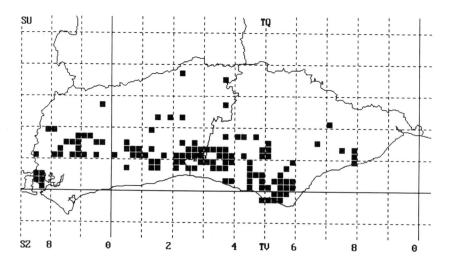

Common Blue

Polyommatus icarus

Common   525 tetrads

Although one of the most widespread of the blues, this butterfly was relatively scarce in 1992 and 94, when many areas contained below average numbers. It is to be hoped that the fine spell at the end of 1994 means that populations will increase once again. Common Blues are found mainly during June and again in August, all along the Downs and also in rough fields and roadside verges throughout the Weald. Many small colonies have doubtless been wiped out by general `tidying-up` and intensive cultivation. We can encourage them to survive even on garden lawns by leaving the main larval food-plant, bird`s-foot trefoil, to spread and by less intensive mowing. They are commonly seen in those churchyards now being managed with wildlife in mind.

The only butterfly in gardens that is likely to be confused with this species is the Holly Blue which has a pale silver-blue underside with small black dots. By contrast the underside of the Common Blue has marginal orange markings and bolder black spots. They also differ in their behaviour. The Common Blue usually flits about at ground level whereas the Holly Blue flies around shrubs and trees and along hedgerows at or above head height.

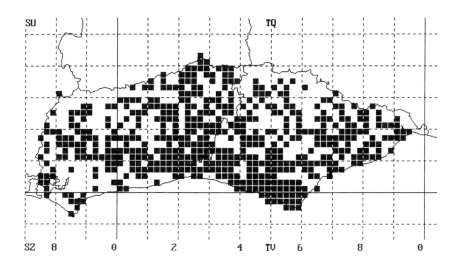

Chalkhill Blue
Lysandra coridon

Locally common   113 tetrads

This is one of our larger blue butterflies, and the male, with its silver-blue upper sides, can be easily recognised flying during July and August over the open downland of both East and West Sussex. The female is more difficult to identify, having brown uppersides and chequered wing margins very similar to the appearance of the female Adonis Blue.

Although the numbers in any one colony do fluctuate from year to year, clouds of these butterflies can occasionally be seen in sheltered warm spots on the Downs. There are fewer colonies than during the 1970s on the western end of the Downs where the larval foodplant, horseshoe vetch, is now much less frequent, but the butterfly has hung on in small numbers at Kingley Vale. At least one small colony has persisted well to the north of the chalk Downs through the six years of the survey and in this case the larval foodplant is thought to be one of the bird`s-foot trefoils.

This butterfly can be seen at Bo-peep Bostal, East Sussex, and at Mill Hill local nature reserve, near Shoreham-by-Sea.

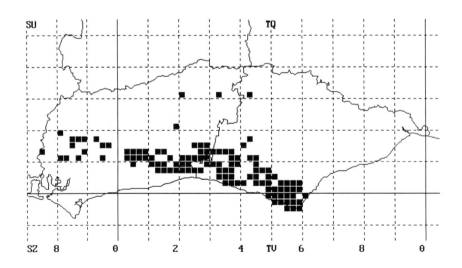

Adonis Blue

Lysandra bellargus

Locally frequent   54 tetrads

Nationally the Adonis Blue is a rare butterfly so we are fortunate that in Sussex it can be found on the warmer grazed slopes of the Downs. Although a few of its colonies are large, many smaller ones survive from year to year. The numbers fluctuate each season eg in May 1993 there were low numbers, and the second brood in August was even poorer, but 1994 was much better with a good late summer brood on many sites.

The male can be recognised by its distinctive upper wings which are a shining brilliant blue, the fore- and hind-wings having a white fringe crossed by black veins (known as chequered margins). The female has similar wing margins, but the upper wing colour is brown -- it is almost identical to the female Chalkhill Blue.

As this butterfly is near the northern edge of its range it has rather specialised requirements. It is essential that its larval food plant, the horseshoe vetch, is not swamped by taller vegetation, and where the grazing is insufficient additional conservation measures are called for.

Adonis Blues can be seen at Castle Hill national nature reserve near Brighton and Mill Hill local nature reserve at Shoreham-by-Sea.

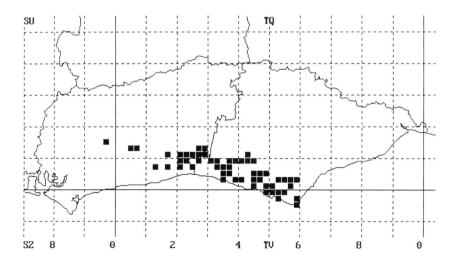

43

Holly Blue
Celastrina argiolus

Common   492 tetrads

During the first years of our survey the Holly Blue was abundant and was widely reported from parks, gardens, hedgerows and from the Downs. However by 1992 few recorders saw any at all, and 1993 was if possible even worse. This is thought to be due to the predations by a parasitic wasp. 1994 saw a slight upturn in numbers and there were sightings in 13 extra tetrads.

This is the blue butterfly most likely to be seen in the garden, flying high at hedge level, basking on leaves or nectaring on blackberry flowers. (In contrast to the Common Blue it does not normally fly close to the ground.) It has a silvery-blue appearance when flying, and when it is at rest it is usually possible to see the small black dots on its grey-blue under surface.

The male is territorial and often parades back and forth along the same stretch of hedge intermittently for several days. The females lay their eggs in May mainly on the flower buds of holly. These develop and emerge as second brood adults in July and August. These second brood females then lay their eggs on the flower buds of ivy or occasionally other shrubs such as dogwood. The larvae can be found by looking for the holes in the flower-buds, and occasionally by the presence of ants crawling in the vicinity.

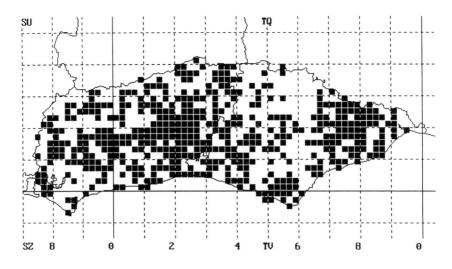

Duke of Burgundy

Hamearis lucina

Rare   22 tetrads

Although often referred to as the Duke of Burgundy Fritillary, it is not a true fritillary despite having rather similar brown and orange wing markings. During our survey it was recorded almost exclusively from the Downs of West Sussex and appears to have gone completely from East Sussex. Several former colonies have been lost due to changes in land use, eg the large colony at Shave's Wood, Poynings, disappeared in the 1980s when conifers were planted.

Many of the existing butterfly colonies are small and very vulnerable, and unless woods are coppiced and scrub kept at bay they risk becoming extinct. The Duke of Burgundy needs sheltered but sunny open clearings with large primrose or cowslip plants on which the female can lay her eggs. West Sussex County Council and the Sussex Downs Conservation Board, with advice and support from our society, have undertaken a great deal of conservation work on the Downs. This has entailed removing scrub to encourage the larval foodplants, primrose and cowslip, and to provide open corridors so that the butterfly can fly to colonise new areas.

The Duke of Burgundy flies during late May and early June and the male is easily seen when it perches on small twigs to defend its territory from any passing butterfly. They usually return time and again to the same place.

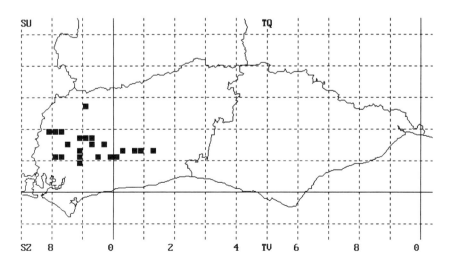

45

White Admiral
Ladoga camilla

Frequent   170 tetrads

    This handsome butterfly has increased in numbers during this century and can be found in most Sussex deciduous woods that have old stands of trailing honeysuckle. The adults are out in late June and July, and can often be seen in ones or twos flying high in the tree canopy or basking on leaves in the sunshine. The woodland around Madehurst in West Sussex is a good place to see them.

    Its deep brown-black upper surface has two wide white bands across both wings. These bands can also be seen on the under surface which has in addition orange-brown areas with black spots and dashes. No other Sussex butterfly has a similar pattern, although when viewed from a distance it can sometimes be mistaken for a Purple Emperor. However the latter is much larger with a more powerful flight, and has eye-spots on the upper surface.

    The young caterpillar stages of the White Admiral can be found up to the end of September by examining the thin dangling shoots of honeysuckle growing in the shade. When the caterpillar emerges from the glassy green egg it eats the leaf until only the midrib is left, and it is this that makes it relatively easy to spot. For much of the day it sits on the projecting remains of the midrib.

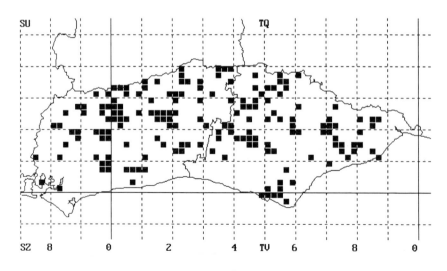

46

Purple Emperor
Apatura iris

Scarce   29 tetrads

    This magnificent butterfly, the largest woodland species in Britain, spends most of its time around the top of the tree canopy, where occasionally one can be fortunate to catch a glimpse of the gleaming purple on the wings that characterises the male. The only butterfly that can be confused with the Purple Emperor and that flies in similar situations is the White Admiral -- this is smaller and the underside pattern is neater.

    Despite being large and conspicuously marked, the Purple Emperor is an elusive butterfly and is probably under-recorded. During our survey it has only been reported from West Sussex despite searches in suitable areas in East Sussex. Low numbers occur in the deciduous woods on the western end of the Downs and in the Low Weald.

    They can be observed most readily in July when they congregate in mating areas around one of the taller trees of the canopy. The females then disperse and seek out sallow bushes on which to lay their eggs. This butterfly is attracted to carrion on the ground where it can be seen feeding, sometimes for long periods. It can usually be seen at Ebernoe Common, a Sussex Wildlife Trust nature reserve in north-west Sussex.

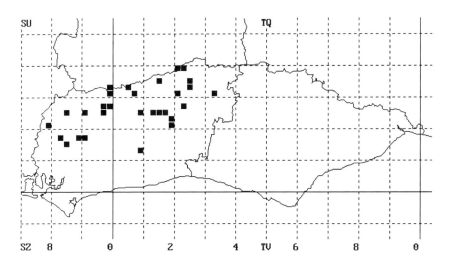

Red Admiral

Common   660 tetrads   *Vanessa atalanta*

    The Red Admiral is a striking butterfly with its rich black wings with their bright crimson markings. It is a yearly migrant from the Continent, but the numbers coming to us are very erratic, 1992 and 1994 being very good years, whereas 1993 was a poor year with far fewer sightings.

    They arrive sometimes as early as March or April and breed here through the summer, becoming one of the commonest and most conspicuous butterflies throughout Sussex. They can be seen right into late autumn, particularly in gardens, although the caterpillars sometimes suffer serious depredations by a parasitic wasp.

    The number of Red Admirals arriving in this country is beyond our control, depending on its breeding success in southern Europe. However, once it is here some of the best ways of helping this species are to leave nettles in a warm situation for it to breed, to grow nectar-rich plants such as buddleia, and in the autumn to leave over-ripe apples for the butterflies to feed on. Despite the recent run of mild winters there has been no definite proof that this butterfly overwinters in Sussex, so records of hibernating adults would be of great interest.

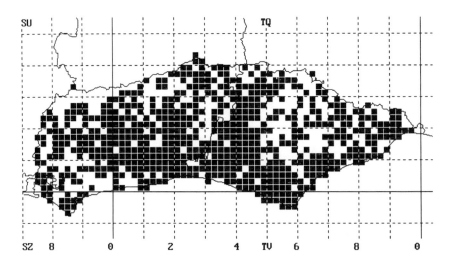

Painted Lady

Cynthia cardui

Frequent   352 tetrads

Another annual migrant from across the Channel, the Painted Lady is most regularly found along the Downs from Beachy Head to the Arundel area. They look pale pinky-orange when flying, but at rest their upper surface of pale orange with black and white markings is unmistakable, They have a fast and powerful flight and have even been seen flying in from the Channel at night. Although not recorded from so many tetrads as the Red Admiral, nevertheless in the occasional good year you can expect to see them almost anywhere in both counties. By contrast 1991, 93 and 94 were poor years with unusually low numbers recorded.

They are unable to survive our winters and the numbers arriving in Britain depend on their breeding success in North Africa and western Europe. Therefore there are only limited practical conservation measures that we can take to increase their numbers here. These are maintaining food nectar sources such as buddleia, and leaving thistles for the caterpillars to feed on.

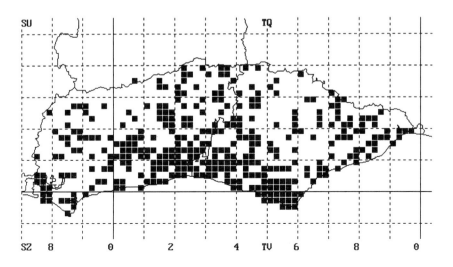

Small Tortoiseshell

Common   686 tetrads   *Aglais urticae*

During most years this is one of our commonest butterflies, and is regularly seen in early spring and in late summer in gardens, along roadside verges and over farmland. However during 1993 and 94 numbers were well down on those of previous years, presumably because of the bad weather, but it is hoped that with a good summer their numbers will soon return to normal.

The female lays large clusters of eggs on young nettles growing in a warm sunny spot. To produce these conditions again for the second brood it is best to cut the nettles in late June so that young fresh shoots can grow in time for the summer females to lay their eggs

As the caterpillars of the Small Tortoiseshell and the Peacock both live gregariously on nettles, they can be easily confused. When they are fully grown, the caterpillars of the Small Tortoiseshell are only about half the size of those of the Peacock, and they are also not so black nor do they appear to be so spiny.

From August onwards the adults nectar eagerly on michaelmas daisies, the ice plant (Sedum) and buddleia, and then in early autumn hibernate in lofts, sheds and cool outbuildings. The underside of both wings is dark mottled brown, providing them with an excellent camouflage, in contrast to the upper surface of bright orange splashed with black and white and bordered with blue dots..

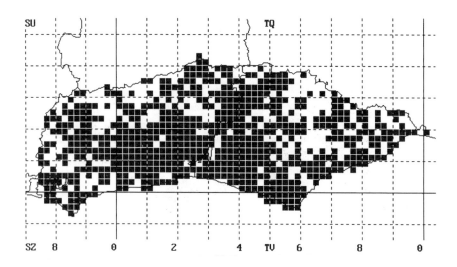

Peacock

Common   632 tetrads

Inachis io

A widely distributed species which hibernates as a butterfly, emerging in the first warm spring days, often as early as February. The deep red-brown upper surface with its four prominent `peacock eyes` contrasts with the very dark, almost black, underside that provides the butterfly with an effective camouflage in its winter hiding places.

In spring it can be found most readily in East Sussex in glades and on the edges of woods although it does occur widely throughout both counties. 1992 was a particularly good year with members reporting large numbers of Peacocks on buddleia bushes throughout late summer and into autumn.

In May the males set up territories and after mating the female lays large batches of eggs on nettles growing in warm sheltered positions eg woodland edges. The caterpillars live together in a web of fine silk and they become jet black and spiny before dispersing to pupate. Although the caterpillars live gregariously on nettles as do those of the Small Tortoiseshell, they are larger and generally suffer a higher mortality through predation by a parasitic wasp.

The adult Peacocks emerge in August and feed avidly on teasels, hemp agrimony, buddleia and michaelmas daisies before going into hibernation..

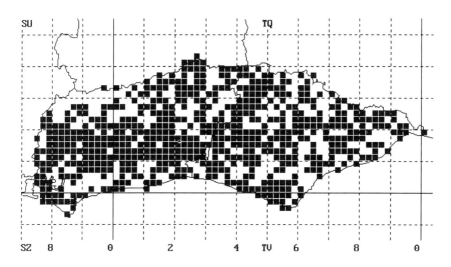

Comma

Common   490 tetrads

Polygonia c-album

The Comma overwinters as an adult and emerges in March or even in February if the weather is mild. The male is often seen perched on a leaf or twig from which he defends his territory, returning to the same place time after time. After mating, the female lays her eggs on nettles, hops or elm. The second brood adults are on the wing from July onwards and are most often seen in ones or twos. In exceptional years such as 1991, up to twelve individuals were counted on one bramble bush in the mid Sussex area.

This is our only butterfly which has a jagged edge to its wings. When in flight they can easily be mistaken for fritillaries, but the underside of the Comma is very different, being mottled brown with a white comma on the hindwing. The upper surface can be either a deep rich orange or a much paler form called *hutchinsoni*.

Commas can be found in East and West Sussex along damp field edges and woodland glades; in Autumn it can also be found in gardens, nectaring on michaelmas daisies or feeding on rotten apples.

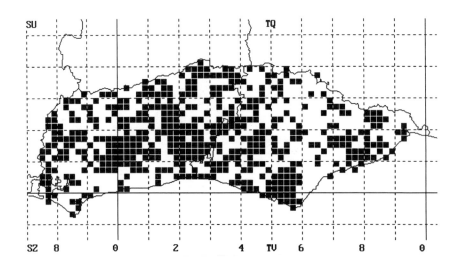

Small Pearl-bordered Fritillary

Boloria selene

Scarce   24 tetrads

In the 1970s this butterfly was known from about 28 localities in West Sussex, but is now down to four tetrads. In East Sussex it is faring a little better and has been found in 20 tetrads since 1989. The Small Pearl-bordered Fritillary is a medium sized fritillary restricted to marshy commons, the early stages of damp coppiced woodland and damp heathland. It is most often seen nectaring on ragged robin and thistles during June. In warm years there can be a second brood, such as there was in 1990, with adults flying during August.

The top surface is orange with black markings but it is the under surface which distinguishes it from the very similar Pearl-bordered Fritillary. Despite their names they are almost the same size. The Small Pearl-bordered Fritillary has an underside with seven silver marginal `pearls` and six or seven additional silvery patches on each hind wing.

The decline of this butterfly is due to the loss of suitable habitat. Conservation measures include retaining damp areas, sunny open woodland rides and ensuring that coppiced areas are large enough. This enables violets, the larval food-plant, to flourish and encourages the growth of various flowers for the adults to nectar on.

They can be seen in small numbers on Chailey Common and on parts of Ashdown Forest, both in East Sussex.

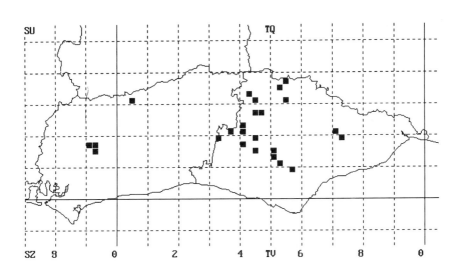

Pearl-bordered Fritillary

Boloria euphrosyne

Scarce   62 tetrads

    This is a rapidly declining species throughout Britain but small colonies can still be found in East and West Sussex especially in newly coppiced woodland or wide sunny rides. It flies from early May to the end of June and is always on the wing at least a fortnight before the Small Pearl-bordered Fritillary. These two species can be distinguished by their undersides -- the Pearl-bordered Fritillary has seven pearls and only two silvery patches (not six or seven) near the base of the hindwing. It also flies close to the ground, whereas the Small Pearl-bordered Fritillary often flies above waist height.

    One of the best places to see the Pearl-bordered Fritillary in West Sussex is at Rewell Wood near Arundel, and in East Sussex in the Vert Wood complex. In both of these areas the woodland is still actively managed. This provides the conditions for violets, the larval food plant, to thrive and also encourages the growth of bugle and other spring flowers that the adult nectars on.

    Coppicing old woodland can encourage the reappearance of these butterflies. In 1989 the BTCV began a programme of woodland management at Lodge Copse near Barlavington, West Sussex. Within a few years the violets colonised the bare ground and the Pearl-bordered Fritillaries were seen again in the wood.

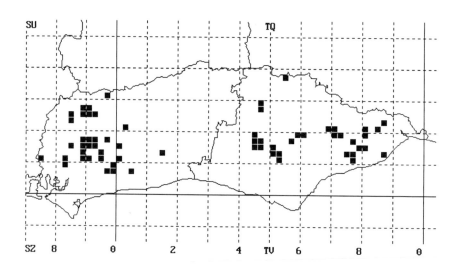

Dark Green Fritillary

Argynnis aglaja

Scarce   65 tetrads

The Dark Green is a large handsome fritillary, with orange wings marked with black on the upper surface. It is similar to the Silver-washed Fritillary but can be distinguished from it by having silver patches on the greenish background of the underside of its hind wings.

It is now only found in low numbers, most of its sites being on rough grassland on the Downs, although a few small woodland colonies have survived in West Sussex. 1990 was a good year for this species and as many as eight were counted one evening at a site near Fulking. By contrast 1992 and 93 were poor years with very few sightings reported.

It can tolerate light grazing preferring a sward of about 6 inches (15cm) that allows its food-plant, the hairy violet, to grow into large clumps. These are easily shaded out when ungrazed grassland becomes tall and rank. Successful conservation measures were started at Cissbury Ring, West Sussex when in 1993 the National Trust re-introduced grazing, leading to a dramatic increase in numbers of the butterfly.

In East Sussex there is a long-standing colony of the Dark Green Fritillary at Birling Gap, near Eastbourne. It flies from the end of June until mid August and because it has a swift and powerful flight, it can be most easily seen when it settles to nectar on such flowers as thistles and knapweed.

Nationally numbers of this fritillary are declining, so records of the sightings of even single individuals are very important.

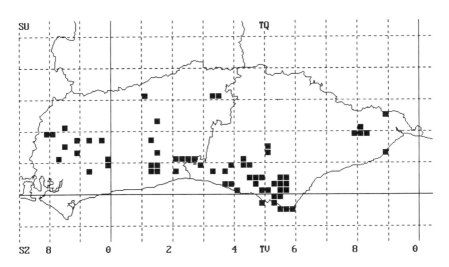

Silver-washed Fritillary

Argynnis paphia

Frequent   170 tetrads

This large butterfly is the fritillary most likely to be seen in Sussex deciduous woodlands from June to late August. The male's bright orange top surface, boldly marked with black spots and streaks, is very noticeable as it often rests on leaves with its wings outstretched in the sunshine. The female is similar but the background colour is a greeny-orange. In both sexes the underside of the hind-wings is a wash of pale silver and green.

The Silver-washed Fritillary has a powerful and gliding flight and can often be glimpsed high amongst the branches. Good views of it can usually be obtained when it comes to nectar on blackberry, thistles or knapweed flowers. It can tolerate shady conditions so has not declined to the same extent as the other woodland fritillaries. Ideally it needs woods with dappled shade and wide sunny rides where there are good patches of dog's violet, the caterpillar's foodplant, and lots of flowers for the adult to nectar.

This fritillary can be seen in many West Sussex woods such as Rewell Wood, near Arundel and Ebernoe Common, north of Petworth, but also is in good numbers in East Sussex in woods near Darwell Reservoir and in the Vert Wood area near East Hoathly.

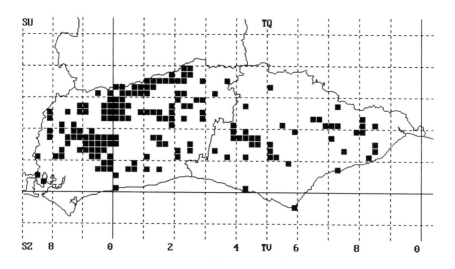

Speckled Wood

Pararge aegeria

Abundant   738 tetrads

A familiar sight, a brown and yellow medium-sized butterfly, fluttering in a shaft of sunlight in a Sussex wood. This is one of the few butterflies you will see in this situation as it can tolerate quite shady conditions compared to other species. The male will defend its small patch of sunlight and will engage in spiral `fights` with other males, returning again and again to the same position.

Although it is generally abundant, numbers fluctuate widely and 1992 was one of those years when sightings were far fewer than usual. This variation in numbers is often out of step with other species. During recent good years it spread beyond its usual woodland habitat and can now be found amongst encroaching scrub on the Downs in East and West Sussex.

The Speckled Wood can hibernate both as a pupa and as a caterpillar which should give it some advantages in the battle for survival. The caterpillar feeds on cocksfoot and couch grass, both of them very common. Adults can be seen almost continuously from late March until October.

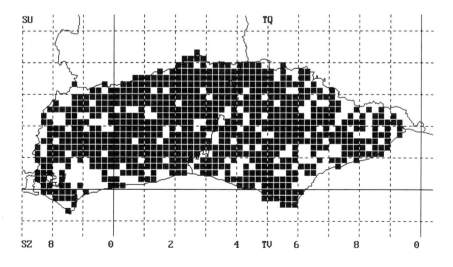

Wall Brown
Locally frequent   278 tetrads

Lasiommata megera.

The Wall Brown is now most frequently seen on the bare sides of downland tracks. It is strongly territorial, defending its stretch of ground and tantalisingly flying ahead as one approaches, settling, then flying on. The upper surface of bright orange and brown markings with conspicuous black rings gives a superficial impression of a fritillary for which it can easily be mistaken when it is flying. When it is at rest with its wings closed the fawn patterns on the undersurface of the hindwing are distinctive.

During the hot year of 1990 very large numbers of the Wall Brown were seen along the Downs, especially north of Shoreham, but since then numbers have been well down, both for the first brood in May and the second in August. The caterpillar can use a range of grasses as its food-plant -- tor grass, false brome, cocksfoot and Yorkshire fog.

Throughout the Weald the Wall Brown seems to be declining, possibly as a consequence of agricultural improvements to the rougher grasslands and field edges where its food-plants are characteristically found, although occasionally it is still to be seen basking there on the sides of dusty arable fields, roadsides and rough pastures. Although in the 1970s it was reliably recorded in the north of West Sussex it is now seldom seen in the north of either East or West Sussex. We are told that it was reported in Surrey from only one site during 1994.

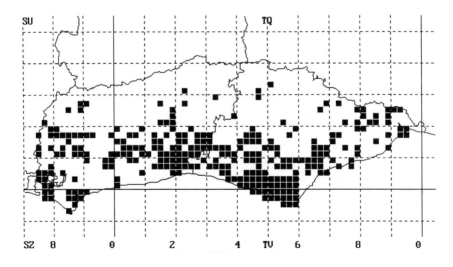

Marbled White

Melanargia galathea

Locally common   146 tetrads

Although the upper surface of the Marbled White has black markings on a white background and superficially resembles one of our 'whites', it belongs to the same family (the 'browns') as the Meadow Brown and Gatekeeper. During July good numbers can often be seen on the Downs in East Sussex, but there are comparatively few large colonies on the West Sussex Downs. Some small colonies survive away from the Downs in unimproved pastures, along woodland rides and even on the verges of a main road in West Sussex, and we have found that these persist from year to year. They are almost certainly remnants of former widespread populations that occurred before farming became intensive and grassland was 'improved', and not, as is commonly supposed, blown off the Downs.

The larvae apparently feed on several types of grass, but for breeding to succeed the grassland should be ungrazed or only lightly grazed and the grasses allowed to grow to their full height. The adults need nectaring plants such as thistles, knapweed and marjoram.

This species can be seen, sometimes in large numbers, at Newhaven Heights and around the Birling Gap area, both in East Sussex.

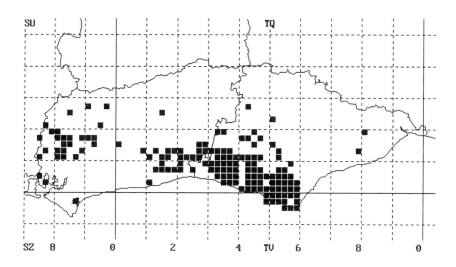

59

Grayling
Hipparchia semele

Rare   15 tetrads

The Grayling is a scarce butterfly in Sussex and the main colony is confined to the Downs of East Sussex. It used to occur on the Greensand around Midhurst in the 1970s but has not been recorded at all in that area during our survey. The occasional individual has been reported on the Downs in the Heyshott area. As these sightings are near the position of records taken in the 1970s there is a chance that small colonies of Grayling still persist in those areas. Its caterpillar food-plant, sheep's fescue, is much less common on the downs than it was, as it is a species easily shaded out by the rank growth of ungrazed grasslands.

In Sussex they are confined to grasslands where there are bare patches in the short turf, on the eroded sides of downland tracks and even in old chalk quarries. Its principal site is a warm valley grazed in winter by cattle whose hoofprints produce the bare patches they like.

The Grayling flies from mid July to early September but is very difficult to spot. It seems to be reluctant to fly, and when it does it goes only a short distance. It then settles on bare areas with its wings tightly closed and aligned to the sun so they do not cast a shadow. The markings on its under-surface break up its outline, so that it is very well camouflaged. The browny-orange top surface with large black eye-spots is hardly ever seen as the butterfly never opens its wings when at rest.

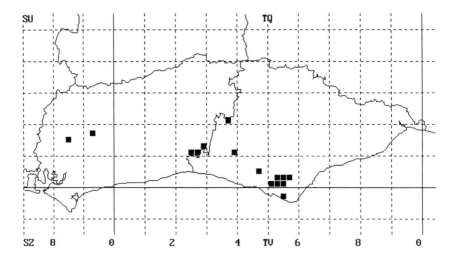

Gatekeeper or Hedge Brown

Pyronia tithonus

Abundant 842 tetrads

This orange and brown butterfly is one of the most abundant in Sussex. It is most typically seen along hedges, woodland rides and increasingly in gardens and wherever there is a patchwork of grass and shrubby cover.

Beginners often find difficulty in distinguishing it from the Meadow Brown. Differences are, the Gatekeeper is smaller and has two white pupils in the black eyespot on the front wing, that it seldom flies over grassland, and that it does not usually settle on the ground. The caterpillar feeds on a wide variety of common grasses, including fescues, cocksfoot, couch and various meadow grasses, which probably helps to explain its wide distribution.

It is on the wing in July and August, although if the season is hot as it was in 1992, the flight period is usually much shorter and few Gatekeepers are seen after mid August. Because it has a short proboscis its nectar sources are more restricted than for many butterflies. It seems to be particularly attracted to flat open yellow flowers such as fleabane and ragwort, but also uses blackberry flowers. On the downs it makes particular use of marjoram, and in the 1991 season large numbers were recorded on this flower above Steyning. It is one of our butterflies that seems to congregate, especially around a good nectar source.

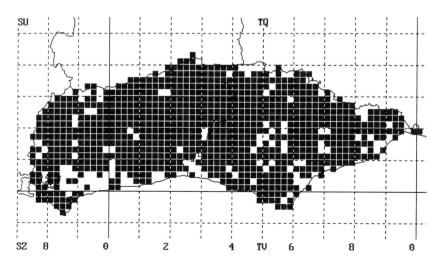

Meadow Brown

Maniola jurtina

Abundant   853 tetrads

This is the familiar brown butterfly most often seen flying over grassland and roadside verges of the Weald, the Downs and the coastal plain. It will breed on any tall wild grass but seems to prefer the various species of meadow-grass and is found in virtually every habitat. During the survey the Meadow Brown was recorded throughout Sussex, although more intensive searching would doutbless reveal further sites. It is on the wing from late May to September with the peak flight period in July and August. Despite this long flight time it has only one brood each year. During the 1990-91 seasons large numbers of Meadow Browns were often reported nectaring on marjoram and knapweed. On the other hand 1993 was a poor year but fortunately during 1994 there was an up-turn in numbers.

In some situations this species can be confused with the Gatekeeper. The Meadow Brown normally has one white pupil in the black eye-spot of the forewing, has a grey-brown appearance, and is larger than the Gatekeeper which always has two white pupils in the black eye-spot, and is a reddish brown.

The Meadow Brown is one of the few butterflies to be seen on the wing during dull, overcast conditions, and has even been known to fly in light rain.

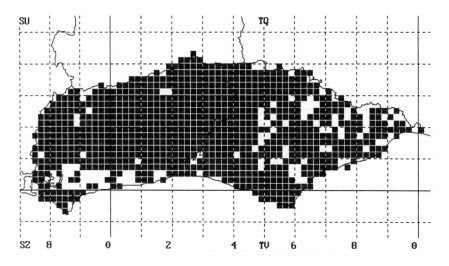

Ringlet

Aphantopus hyperantus

Frequent   360 tetrads

    This attractive dark velvety-brown butterfly has white eyespots encircled by yellow and black rings on the under-wings. When only the upper-surface of rather worn specimens is seen, the male in particular can be confused with the male Meadow Brown but this always has some orange coloration.

    The Ringlet is found throughout East and West Sussex, mainly in the Weald on damper grassland and woodland edge. It is absent from parts of the East Sussex Downs where the grassland tends to be too short and dry, but it can be found on the more wooded Downs of the far west where there are damper spots for the species to breed. It is also absent from nearly all the coastal area of both counties.

    Its caterpillar appears to feed on a variety of grasses such as tufted hair-grass, couch and smooth meadow-grass.

    It has a flight period of only about three weeks in July to early August. It is a much more restless butterfly than the Meadow Brown, often only taking short 'bouncy' flights from one piece of shade to the next. It even flies in the rain when the weather is warm. When the Ringlet is flying it is possible to see its surprisingly conspicuous white marginal fringe, another way of distinguishing it from the Meadow Brown. Bramble flowers are the favoured nectar source of the Ringlet, and this is probably the best way to see them at rest.

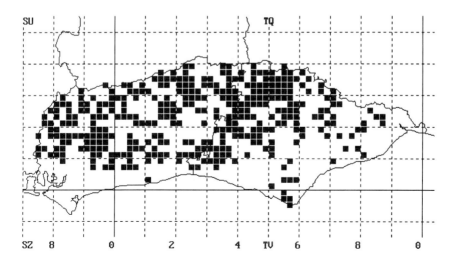

Small Heath

Coenonympha pamphilus

Frequent   337 tetrads

This small butterfly spends most of its time on the ground with its grey-brown wings closed. It tilts its wings towards the sun so does not cast a shadow and is then very inconspicuous. It flies very infrequently, keeping just above the grass, and this is the only time to get a glimpse of its light tawny-orange upper surface. It is easily overlooked especially where the colonies are small in areas like the verges of small lanes in East Sussex.

In the 1970s it was reported to be as common in West Sussex as the Meadow Brown. However, during our 1989-94 survey numbers dropped progressively, reflecting the national trend, although good populations can still be found on the Downs of East Sussex. It is difficult to suggest remedial conservation measures because the requirements of the species are not well understood.

Its characteristic habitat is unimproved grassland, and the caterpillar feeds on a range of grasses including the fescues and the bents.

A good place to see it is Lullington Heath National Nature Reserve, East Sussex especially during August when the second brood is flying.

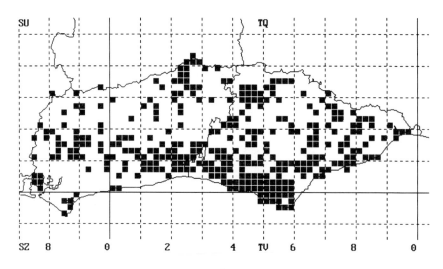

Some notes on 1995

This must surely be remembered as one of the most exceptional butterfly years in recent memory. It started quite well, but gave no early indication of what was to follow.

The record-breaking hot summer seemed to go on and on, and it became clear that several species were going to benefit. It resulted in very good numbers of Chalkhill and Adonis Blues, hardly a surprise as both are species of warm sites. The Holly Blue, as well as appearing to be moving towards a high point in its cycle, produced in September a few new-looking specimens that can only have been those of a third brood.

Other species that in recent years had given cause for concern gave us renewed hope. The Small Tortoiseshell has bounced back, and good numbers of the Dark Green Fritillary were seen on parts of the Downs. The Small Copper, another species that has recently not done well, was reported by one member to be in 'extraordinary numbers' throughout East and West Sussex. The dramatic increase in numbers of several species demonstrated how difficult it can be to identify trends in butterfly numbers at all accurately, and hence to know whether improvements in numbers on a site are 'accidental' or due to changed management.

What really caused great excitment was the influx into Britain of large numbers of the Camberwell Beauty. There were nineteen records from Sussex, many from as far north as Lancashire, and a few from Scotland. This invasion was followed later in the year by thirteen sightings of the Monarch in Sussex, but nationally it did not quite reach the high number of reports of the Camberwell Beauty. The exceptional appearance of these two large beautiful migrants captured the public imagination and even achieved coverage in the national press. The strange thing is that there were so few records of our 'usual' migrant, the Clouded Yellow -- only three in Sussex.

Is all this a taste of things to come?

Part 3   Appendices

Appendix 1

The frequency categories

Abundant

Species which are likely to be seen widely throughout the county, generally in good numbers. They are not particularly narrow in their habitat requirements..

Large White	Small White	Speckled Wood
Gatekeeper	Meadow Brown	

Common

These species are widely distributed through the county but are not seen in sufficient numbers to be regarded as abundant. About half of these are 'wanderers' which usually occur in low numbers at any one site.   Others space themselves out either because they are   territorial with the males defending their patch from other competing males of the same species, or because they form colonies of fairly restricted size.

Small Skipper	Essex Skipper	Large Skipper
Brimstone	Green-veined White	Orange Tip
Common Blue	Holly Blue	Red Admiral
Small Tortoiseshell	Peacock	Comma

Locally Common

These are restricted to certain parts of the county because of their specialised habitat requirements, but wherever they occur they are often so numerous as to form 'clouds'.

Chalkhill Blue	Marbled White

Frequent

These are to be found scattered across the county but are mostly confined to their special habitats. They usually occur in relatively low numbers at any one site. This category brings together species of very different characteristics; with better knowledge some may prove to be more common than is currently thought, whilst others are almost certainly on the decline.

Purple Hairstreak	Small Copper	White Admiral
Painted Lady	Silver-washed Fritillary	
Ringlet	Small Heath	

Locally Frequent

These are generally concentrated in certain parts of the county, eg the Downs, and in good years can occur in quite good numbers, but not sufficient to be regarded as 'clouds'

Dingy skipper	Grizzled Skipper	Green Hairstreak
Silver-studded Blue	Brown Argus	Adonis Blue
Wall Brown		

Scarce

These are found in restricted areas, generally in low numbers.

Brown Hairstreak	White-letter Hairstreak	Small Blue
Purple Emperor	Small Pearl-bordered Fritillary	
Pearl-bordered Fritillary	Dark Green Fritillary	

Rare

Very few sites remain in Sussex.

Silver-spotted Skipper	Wood White	Duke of Burgundy
Grayling		

Casual

Not regularly occurring, numbers and sites varying from year to year depending on migration.

Clouded Yellow

Appendix 2

Butterfly transects in Sussex

There are several sites in Sussex where butterfly transects are being recorded using the standardised method that is recognised nationally, and details of a representative sample of them is given on the opposite page. The method entails walking the same line weekly from April to September every year, noting the numbers of each butterfly species seen on or close to the route of the walk. In this Atlas we used the transect results only insofar as they indicate presence or absence of each species. It should be noted also that the standardised method excludes species that are not on or close to the line of the walk, so the results do not necessarily give a full picture of the butterflies present on the site.

Butterfly transects are essentially concerned with recording changes in the numbers of butterflies on a site from year to year. Transects no.1 to 6 are part of the Butterfly Monitoring Scheme organised nationally by the Institute of Terrestrial Ecology. The purpose of this scheme is to provide a national baseline of the fluctuations in butterfly numbers that local fluctuations in numbers can be judged against. This means that it is possible to see whether, for instance, the decline of a particular species on a particular site is part of a national decline or is just a local phenomenon. The sites included in this national scheme have been chosen to give a good geographical scatter and to represent the range of habitats in Britain. The same sites are recorded each year so that the results of one year can be properly compared with those of another.

Transects no.7 to 11 are not part of the national scheme, but use the same standardised method. By comparing their results with those of the national scheme it is possible to show whether changes from year to year are part of a national trend or are due to local circumstances that would require, for instance, an adjustment to the site management.

Some examples of butterfly transects

1--Castle Hill National Nature Reserve, near Brighton. Chalk grassland.
2--Kingley Vale National Nature Reserve, near Chichester. Chalk grassland.
3--Lullington Heath National Nature Reserve, near Eastbourne. Chalk heath.
4--Morris's Wood, near Crowborough. Private deciduous woodland. Recorded to 1992.
5--West Dean Woods, Sussex Wildlife Trust Reserve near Chichester. Hazel coppice.
6--Woods Mill, Sussex Wildlife Trust Reserve, near Henfield. Deciduous woodland and meadow.
7--Ebernoe Common, Sussex Wildlife Trust reserve, north of Petworth. Ancient deciduous woodland .
8--Levin Down, Sussex Wildlife Trust Reserve, near Chichester. Chalk grassland and chalk heath.
9--Malling Down, Sussex Wildlife Trust Reserve, near Lewes. Chalk grassland.
10-Rye Harbour Local Nature Reserve and Sussex Wildlife Trust Reserve, near Rye. Shingle and rough fields.
11-Thorney Island, near Chichester. Coastal

Appendix 3

Check list of caterpillar food-plants referred to in the text

The scientific names follow
CLAPHAM A R TUTIN T G & MOORE D M
Flora of the British Isles Cambridge (1987)

Alder buckthorn	*Frangula alnus*
Bent grass	*Agrostis* spp
Bird's-foot trefoil	*Lotus corniculatus*
Bitter vetch	*Lathyrus montanus*
Blackthorn	*Prunus spinosa*
Cabbage	*Brassica* cultivars
Clover	*Trifolium* spp
Cocksfoot	*Dactylis glomerata*
Common sorrel	*Rumex acetosa*
Common storksbill	*Erodium cicutarium*
Couch grass	*Elymus repens*
Creeping cinquefoil	*Potentilla reptans*
Creeping soft-grass	*Holcus mollis*
Cowslip	*Primula veris*
Dog's violet	*Viola riviniana*
Dogwood	*Cornus sanguinea*
Dove's-foot cranesbill	*Geranium molle*
Elm	*Ulmus minor*
False brome	*Brachypodium sylvaticum*
Fescue	*Festuca* spp
Garlic mustard	*Alliaria petiolata*
Gorse	*Ulex europaeus*
Hairy violet	*Viola hirta*
Heath	*Erica* spp
Heather	*Calluna vulgaris*
Hedge mustard	*Sisymbrium officinale*
Holly	*Ilex aquifolium*
Honesty	*Lunaria annua*
Honeysuckle	*Lonicera periclymenum*
Hop	*Humulus lupulus*
Horseshoe vetch	*Hippocrepis comosa*

Ivy	*Hedera helix*
Kidney vetch	*Anthyllis vulneraria*
Lady's smock	*Cardamine pratensis*
Lucerne	*Medicago sativa*
Meadow-grass	*Poa* spp
Meadow vetchling	*Lathyrus pratensis*
Nasturtium	*Tropaeolum* spp
Nettle	*Urtica dioica*
Oak	*Quercus* spp
Primrose	*Primula vulgaris*
Purging buckthorn	*Rhamnus catharticus*
Rockrose	*Helianthemum nummularium*
Sallow	*Salix caprea* and *S. cinerea*
Sheep's fescue	*Festuca ovina*
Sheep's sorrel	*Rumex acetosella*
Smooth meadow-grass	*Poa pratensis*
Thistle	*Carduus* spp and *Cirsium* spp
Tor grass	*Brachypodium pinnatum*
Tufted hair-grass	*Deschampsia cespitosa*
Violet	*Viola* spp
Wild strawberry	*Fragaria vesca*
Wych elm	*Ulmus glabra*
Yorkshire fog	*Holcus lanatus*

Appendix 4

Some useful books

The following is a selection of the books currently available.

THOMAS J A
> **Butterflies of the British Isles**   (Hamlyn Guide)
> Hamlyn **1992**
> Gives a two page spread on each species, information on flight periods and an identification chart. Widely regarded as the most useful single book.

CHINERY M & HARGREAVES B
> **Butterflies and Moths** (Collins Gem)
> Harper Collins **1981**
> A useful pocket-sized book for the beginner

WHALLEY P
> **The Pocket Guide to Butterflies**
> Mitchell Beazley **1981** and later editions
> Covers Britain and Continental Europe, with illustrations and brief notes of almost every species.   The slim format makes it an excellent book for the pocket.

CHINERY M
> **Butterflies and Day-flying Moths of Britain and Europe**
> (New Generation Guide)
> Collins **1989**
> As well as illustrations and brief descriptions of the butterflies and many moths, it gives small maps of their distribution in Western Europe. It also has a large section on the biology of butterflies and moths, including such topics as pests and migration. Comprehensive, but rather large for the pocket.

THOMAS J A & LEWINGTON R
> **The Butterflies of Britain and Ireland**
> Dorling Kindersley **1991**
> Not a book for the field, but one that is full of up-to-date information on all aspects of the natural history of British and Irish species. Beautifully illustrated.

HYDE G E

Butterflies Sticker Book   (Usborne Spotter's Guides)
Usborne **1944**
A book to kindle children's interest, and its stickers are some
of the best butterfly illustrations available.

EMMET A M & HEATH J (Editors)
The Moths and Butterflies of Great Britain and Ireland
Vol. 7, Pt 1 Hesperiidae-Nymphalidae: The Butterflies
Harley Books **1989**
The authoritative and comprehensive book on British
butterflies. Essentially a book for the specialist and dedicated
amateur.

PRATT C
A History of the Butterflies and Moths of Sussex
Booth Museum, Brighton **1981**
A detailed account of the distribution of each species of Sussex
butterfly up to 1979, with maps of the scarcer species.   A
valuable reference work.

BUTTERFLY CONSERVATION
publishes occasional booklets on individual species (such as
the Purple Emperor) and topics (such as gardening   to
encourage butterflies).   Articles on butterflies, their ecology
and conservation appear in the society's **News** which is
published three times a year and is free to members. Up-to-
date information can be obtained by writing to the society at
PO Box 222, Dedham, Colchester CO7 6EY.

ENGLISH NATURE
also publishes booklets on species, habitats and their
conservation (such as one on **Heathland Management for the
Silver-studded Blue Butterfly**). Write for an up-to-date list
to Northminster House, Peterborough PE1 1UA.

HALL P C
Sussex Plant Atlas
Booth Museum, Brighton **1980**
A detailed account of the distribution of flowering plants in
Sussex up to 1978, with tetrad maps of most species. The
Selected Supplement published in **1990** gives additional
records to 1988.

Appendix 5

Where and when to see butterflies in Sussex

Home and around

One of the best places to see butterflies can be in your own garden or local park. If you live in town and have a large varied garden with a wide variety of plants and flowers you are likely to see up to 20 species of butterfly in a good year. These could include six species of the whites, three skippers, both the Holly and Common Blues, several of the larger species (the vanessids) such as Small Tortoiseshell, Red Admiral, Peacock and Painted Lady, and three browns, Speckled Wood, Gatekeeper and Meadow Brown. Even a small garden can attract at least 12 species, especially if you make sure you grow garden flowers that have a strong scent and lots of nectar. Garden flowers do attract the larger butterflies from quite a distance, but you can, so to speak, 'grow your own' by leaving patches of nettles and wild plants like lady's smock. Leave holly in the hedge and let ivy grow large enough to flower if you want to encourage Holly Blues. Bramble flowers are attractive to many butterflies particularly in August. The enthusiast can even plant buckthorn to encourage the Brimstone to breed in the garden. Several butterfly books give helpful hints on how to encourage butterflies in your garden -- see the booklist in Appendix 4.

It is also worth exploring areas close to home, just on the edge of town. Places like Bedelands on the edge of Burgess Hill, Southwater Country Park just south of Horsham, and Marline and Park Woods nature reserve on the outskirts of Hastings are some typical examples of the sort of places that are often overlooked but have proved to be good for butterflies.

Species such as the Small and Large Skippers, the Small Copper and Common Blue are likely to turn up on almost any rough grassland and roadside verges on the smaller side-roads can be very rewarding for these and others of the commoner butterflies. Early in the year many of them are good places to see the Orange Tip. Nowadays many churchyards have sections that are managed to encourage their wildlife, and these are often good for a range of butterflies -- as a bonus several are notable for their orchids.

The Downs

To see the typical downland butterflies visit one of the county's nature reserves or National Trust areas during the summer months.. All of these allow public access, at the very least along the public footpaths, and many have car parks. Harting Down in the far west of the county, Mill Hill near Shoreham-by-Sea, Ditchling Beacon, the Seven Sisters Country Park near Seaford and the Beachy Head area are well worth exploring.

There is also easy access to many parts of the Downs along the numerous public footpaths and bridleways, and many butterflies can be seen especially where the original downland vegetation has been left.

Heathland

To see the heathlands species such as Silver-studded Blue choose from Iping Common and the immediately adjoining Stedham Common Local Nature Reserves near Midhurst in West Sussex, or wander along regularly mown firebreaks of Ashdown Forest in East Sussex. In general there are fewer species to be seen on these more acid and sandy areas than there are on the Downs, but they do have different species, so take advantage of the great variety that we have in Sussex as a result of our geological diversity.

Woods

A good range of the woodland species can be seen at such places as Rewell Wood near Arundel and the Sussex Wildlife Trust's Ebernoe Common nature reserve north of Petworth -- a bit difficult to find but a fine example of ancient woodland that is well worth the effort. The Vert Wood area near East Hoathly and the Beckley Woods and Flatropers Wood near Rye also have a good range of the characteristic woodland species.

The species to look out for in Summer are the Silver-washed Fritillary and White Admiral floating down into the glades, Speckled Wood should be in many of the sunny patches. If you are really lucky in May there will be Pearl-bordered Fritillaries, and later in the year, Purple Hairstreaks. The Brimstone will have made its conspicuous appearance in early spring and again from late summer onwards, especially in damper woods where buckthorn can grow well. Most Forest Enterprise areas have a wide selection of woodland butterflies wherever there are wide rides, especially when deciduous trees have been retained along the ride margins.

When?

To improve your chances of seeing butterflies, choose a warm sunny day from mid-morning onwards, and check the accompanying chart to see if the butterfly you want to see is flying in that season. It is no good trying to see Chalkhill Blues in May ! If the weather is very hot butterflies can be on the wing very early in the day and can be seen again towards evening when many are feeding. At mid-day in hot spells many species go into hiding.

The chart below showing approximate flight times through the season has been prepared on the basis of records from Sussex, and the beginnings and endings of the flight seasons may differ slightly from those given in books about Britain as a whole. Also in hot summers species tend to fly earlier in the year than usual, so take account of this when consulting the chart. Early and late in the year (outside the months in the chart) the hibernating species often appear on the occasional warm day, especially when they are accidentally disturbed.

Flight times of Sussex Butterflies

```
Skippers       Apr     May     Jun     Jul     Aug     Sep
Sm Skipper                          ++++++++++++++++++
Esx Skipper                             ++++++++++++++++
Sil-spd Skipp                              +++++++++++++
Lge Skipper                        +++++++++++++++++++++
Dingy Skipp          ++++++++++++++++++             ++
Grizz Skipper    +++++++++++++++                    +

Whites         Apr     May     Jun     Jul     Aug     Sep
Wood White           ++++++++++                  ++
Clouded Yellow                      +++++++++++++++++++++ + +
Brimstone        +++++++++++++++++++                +++++++++
Large White        ++++++++++++++++++++++++++++++++++++++++++++++
Small White      +++++++++++++++++++       ++++++++++++++++++
Gr-vnd White     ++++++++++++++++++++++++      ++++++++++++
Orange Tip       +++++++++++++++++

Hairstreaks and
Small Copper   Apr     May     Jun     Jul     Aug     Sep
Green Hairstk        ++++++++++++
Brown Hairstk                             ++++++++++++++
Purple Hairstk                       +++++++++++
Wh-let Hairstk                       +++++++
Small Copper     ++++++++++++            ++++++++++++++++ +
```

Blues	Apr	May	Jun	Jul	Aug	Sep

Blues Apr May Jun Jul Aug Sep
Small Blue   +++++++++   +
Silr-std Blue   +++++++++++
Brown Argus   ++++++++++++++++++   +++++++++++++++++
Common Blue   +++++++++++++   +++++++++++++++++
Chalkhill Blue   ++++++++++++
Adonis Blue   ++++++++++   ++++++++
Holly Blue +++++++++++   ++++++++++++++++

Duke of Burgundy   +++++++

Emperor, Admirals
and Vanessids Apr May Jun Jul Aug Sep
White Admiral   ++++++++++++
Purple Emperor   +++++++++++++++
Red Admiral   ++
Painted Lady   + + + + + + + +++++++++++++
Small Tortshell +++++++++++++   +++++++++++++ ++++++++
Peacock +++++++++++++   ++++++++++++++++
Comma ++++++++++   ++++++++++++++++++++

Fritillaries Apr May Jun Jul Aug Sep
Sm P-bd Frit   +++++++++++   +
Pearl-bd Frit   ++++++++++   +
Dk Grn Frit   ++++++++++++++++
Sil-wshd Frit   ++++++++++++++++++

Browns Apr May Jun Jul Aug Sep
Speckd Wd +++ + +
Wall Brown   ++++++++++   ++++++++++++++
Marb White   ++++++++++++++++
Grayling   +++++++++++++
Gatekeeper   ++++++++++++++++
Mdow Brown   ++++++++++++++++++++++++
Small Heath   +++++++++++++++++   +++++++++++
Ringlet   +++++++++

Notes

Acknowledgements

--Jim Steedman for help in so many ways, especially on all aspects of computerisation of the records, without which the survey could not have been completed so quickly. He devised the computer program that formed the basis of our record system, and spent many hours patiently helping us over our difficulties

--Thomas Barber for his cover illustration

--Colin Pratt for providing various information

--Susan Rowland of the Geography Laboratory of the University of Sussex for providing the special geology map

--Colin Simmonds for reading the draft text and for providing such characteristically helpful suggestions

--East and West Sussex County Councils, especially for financial support of various parts of the whole project

--English Nature for financial support towards publication of this Atlas

--Last but not least **the volunteers.** By their persistence they have provided the records that are the essential core of this Atlas. Deserving of special mention are

N Armfelt, H Avery, G Ayres, T Baigent, P Bailey, J Baker, D Bangs, I Barber,
J Barbour, T Barber, C Barley, C Barnard, D & E Bashford, I Beavis, R & M Becker,
V Bentley, C Best, M Biggs, G & B Bishop, P Blackwell, L Bowman, J Brackpool,
P Brown, E Bruce, R Bruce, D Buckingham, D Burrows, F Carson, P Catlett,
Richard Carter, Roger Carter, D Chambers, G & M Champion, C Clarkson Webb,
P Clarkson Webb, J Claydon, C B Collins, S Comerford, S & T Crabbe, R Crane,
K Crook, S Curson, D Dancy, H Davies, P Davys, A de Potier, M de Saulles, D Dey, B Dorey,
P Dunn, W Ellis, M Emberson, M Ewing, P Farrant, J Finch, J Franks, E & C French,
B Frost, D Funnell, D Gaydon, J Gent, A Gillham, J Good, R Gould, M Green, A Greenwood,
A Griffiths, A Guest, J Gumbrell, J Halls, S Hamilton, S Hanson, D Harris, P Harrison,
G Hart, J Havers, K Hearne, D Henderson, B Hey, J Hicks, D Hill, R Hinks, D Hobden,
A Holden, J Holloway, L Holloway, M Holmes, P Howes, K Hurle, F Jackson, G Jeffcoate,
D Jupp, S Kellaway, R Kemp, D Knapman, S Knapp, J Landan, M Le Grys, I Lewis, D King,
T Lavender, A Low, U Lutwych, J Maddocks, A Malpass, S Manning, L Marson, D Maynard,
R Meredith, D Middleton, N Midlane, N Mitchell, E Mockford, J Morley, P Mottershead,
R Neeve, G Parris, S Patton, J Payne, R Pepper, J Phillips, S Poole, C Pratt, C Price,
D & K Pritty, B H Reed, B & D Reeve, E & K Reeves, G Revin, N Rhys-Williams,
C & P Rhodes, G Roberts, P Roper, P Rowland, J Rudd-Jones, A Rush, D Russell, M Russell,
D Sadler, W Salkeld, M Senior, J Seymour, J & R Sharpe, J Shaughnessy, C Simmonds,
A Simpson, J & J Steedman, A Steele, D Steele, P Stuart, C Slack, B Stoner, J Syms,
B Taylor, S Thomas, M Trew, K Trillwood, L Turton, C Wagstaff, J Wagstaff, A Wedd,
J Wells, A Went, A Whitbread, P Whitcomb, S White, K & J Whiteman, A Widdicombe,
B Wilkins, A & R Williamson, W Wisden, J Woodhouse

Do you want to see more butterflies in Sussex?

One of the most enjoyable ways of seeing butterflies is to join with others who have similar interests. Butterfly Conservation is a registered charity dedicated to saving wild butterflies and their habitats, and which through its Sussex branch organises a whole range of activities in Sussex. Its programme includes weekly summer field visits to see butterflies, winter social evenings where members can view each other's slides, and talks by specialists. We spread our activities throughout the county so that all members can take part more easily. You can also help with practical conservation tasks, and contribute your records to the continuing up-dating of our records -- this Atlas was only possible because of such help from so many members.

We give a lot of effort to persuading private landowners and public bodies to take fuller account of the needs of butterflies. In Sussex our society owns and manages a small reserve specially for butterflies at Park Corner Heath near East Hoathly which is open for visits by members. It is a site where White Admirals and Silver-washed Fritillaries can be seen in profusion. It is managed almost entirely by the help of our members and other volunteers.

To join Butterfly Conservation and the Sussex branch

or for more information write to

P O Box 222, Dedham, Colchester, Essex, CO7 6EY.

Some public libraries in Sussex also have leaflets about the Sussex branch.